Editor
Lorin Klistoff, M.A.

Managing Editor
Karen Goldfluss, M.S. Ed.

Editor-in-Chief
Sharon Coan, M.S. Ed.

Cover Artist
Barb Lorseyedi

Art Coordinator
Kevin Barnes

Art Director
CJae Froshay

Imaging
Alfred Lau
James Edward Grace
Rosa C. See

Product Manager
Phil Garcia

Publisher
Mary D. Smith, M.S. Ed.

Author
Mary Rosenberg

Teacher Created Resources, Inc.
6421 Industry Way
Westminster, CA 92683
www.teachercreated.com

ISBN: 978-0-7439-3313-1

Reprinted, 2014
Made in U.S.A.

Table of Contents

Introduction

The old adage "practice makes perfect" can really hold true for your child and his or her education. The more practice and exposure your child has with concepts being taught in school, the more success he or she is likely to find. For many parents, knowing how to help their children can be frustrating because the resources may not be readily available. As a parent it is also difficult to know where to focus your efforts so that the extra practice your child receives at home supports what he or she is learning in school.

This book has been designed to help parents and teachers reinforce basic skills with their children. *Practice Makes Perfect* reviews basic math skills for children in grade 3. The math focus is word problems. While it would be impossible to include in this book all concepts taught in grade 3, the following basic objectives are reinforced through practice exercises. These objectives support math standards established on a district, state, or national level. (Refer to the Table of Contents for the specific objectives of each practice page.)

- rounding to the nearest ten and hundred
- adding without regrouping
- adding with regrouping
- dividing with remainders
- adding and subtracting money
- using probability and greater than/less than
- subtracting without regrouping
- subtracting with regrouping
- multiplying and dividing
- using time
- adding and subtracting fractions
- finding the area and average

There are 36 practice pages organized sequentially, so children can build their knowledge from more basic skills to higher-level math skills. To correct the practice pages in this book, use the answer key provided on pages 47 and 48. Six practice tests follow the practice pages. These provide children with multiple-choice test items to help prepare them for standardized tests administered in schools. As children complete a problem, they fill in the correct letter among the answer choices. An optional "bubble-in" answer sheet has also been provided on page 46. This answer sheet is similar to those found on standardized tests. As your child completes each test, he or she can fill in the correct bubbles on the answer sheet.

How to Make the Most of This Book

Here are some useful ideas for optimizing the practice pages in this book.

- Set aside a specific place in your home to work on the practice pages. Keep it neat and tidy with materials on hand.
- Set up a certain time of day to work on the practice pages. This will establish consistency. An alternative is to look for times in your day or week that are less hectic and more conducive to practicing skills.
- Keep all practice sessions with your child positive and constructive. If the mood becomes tense or you and your child are frustrated, set the book aside and look for another time to practice with your child.
- Help with instructions if necessary. If your child is having difficulty understanding what to do or how to get started, work the first problem through with him or her.
- Review the work your child has done. This serves as reinforcement and provides further practice.
- Allow your child to use whatever writing instruments he or she prefers. For example, colored pencils can add variety and pleasure to drill work.
- Pay attention to the areas in which your child has the most difficulty. Provide extra guidance and exercises in those areas. Allowing children to use drawings and manipulatives, such as coins, tiles, game markers, or flash cards, can help them grasp difficult concepts more easily.
- Look for ways to make real-life application to the skills being reinforced.

Practice 1

Round each number to the nearest ten. Then add or subtract.

- If the number in the ones place is 5 or larger, round up. Example: 4<u>6</u> → 50
- If the number in the ones place is 4 or less, round down. Example: 4<u>4</u> → 40

1. Courtney had 18 red jelly beans and 41 green jelly beans. About how many jelly beans did she have in all?

   Courtney had about _____ jelly beans in all.

2. Orville had 62 points. Oliver scored 17 more points. About how many points were scored in all?

   About _____ points were scored in all.

3. Asia found 53 seashells and 83 starfish. About how many more starfish did Asia find than seashells?

   Asia found about _____ more starfish.

4. Sully picked 14 berries and 35 plums. About how many more plums did Sully pick than berries?

   Sully picked about _____ more plums.

5. Brandy had 89 cows and 3 goats. About how many animals did Brandy have in all?

   Brandy had about _____ animals.

6. Tim gathered 17 rocks and 46 small pebbles. About how many more pebbles did Tim gather than rocks?

   Tim gathered about _____ more pebbles.

Practice 2

Remember the following facts when rounding numbers to the nearest hundred:

- If the number in the tens place is 5 or larger, round up. Example: 3<u>9</u>1 → 400
- If the number in the tens place is 4 or less, round down. Example: 3<u>1</u>1 → 300

Round each number in the chart to the nearest hundred.

	Chin Ups	Push Ups	Sit Ups
Callie	63	110	105
Brynn	51	192	212
Melvin	87	181	210
Paz	74	187	159

Use the rounded numbers above to answer each question. Fill in the correct answer circle.

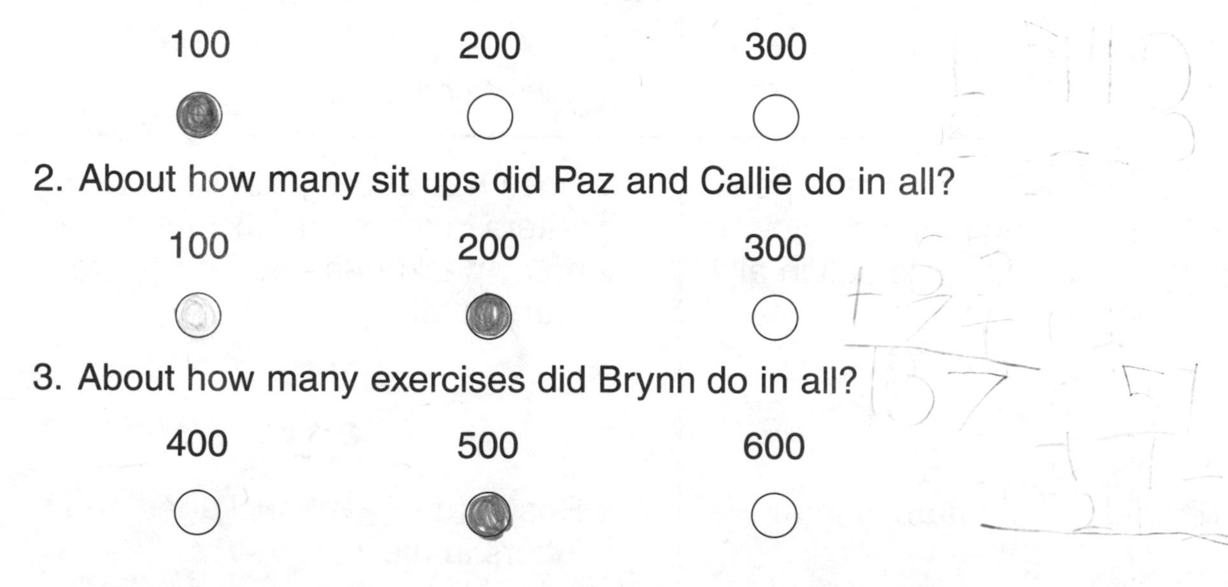

1. About how many chin ups did Brynn and Melvin do in all?

   100 ○ 200 ○ 300 ○

2. About how many sit ups did Paz and Callie do in all?

   100 ○ 200 ○ 300 ○

3. About how many exercises did Brynn do in all?

   400 ○ 500 ○ 600 ○

Practice 3

Solve each word problem.

1. Joselyn delivers mail to 865 homes. So far, she has delivered mail to 423 homes. How many more homes does Joselyn still need to deliver mail to?

$$\begin{array}{r} 865 \\ -\ 423 \\ \hline \end{array}$$

Joselyn needs to deliver mail to ______ more homes.

2. Rocky collects postcards. His aunt sent him 415 postcards while on her trip to Egypt. His uncle sent him 251 postcards while on his trip to Germany. How many postcards did Rocky receive?

$$\begin{array}{r} 415 \\ +\ 251 \\ \hline \end{array}$$

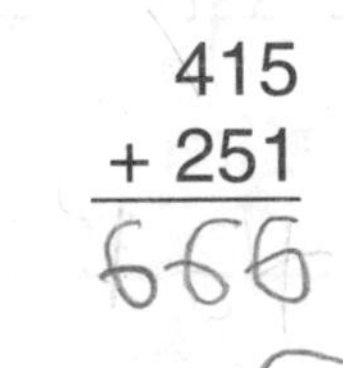

Rocky received ______ postcards.

3. Grace had 757 pieces of mail. 344 of the pieces were postcards. The rest were letters. How many letters did Grace have?

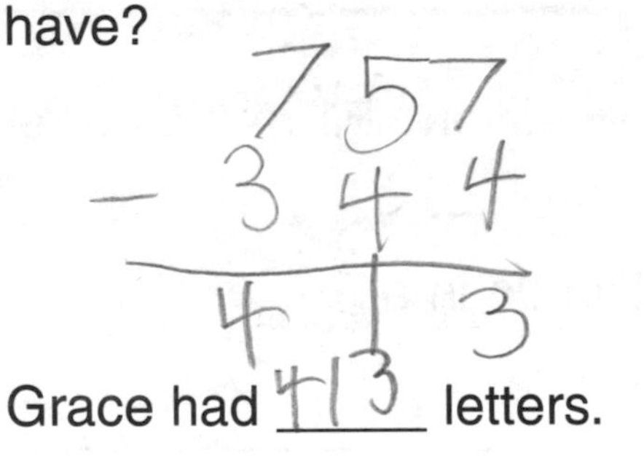

Grace had ______ letters.

4. Tony delivered 423 packages. 101 of the packages were stamped "one-day delivery." The rest were stamped "two-day delivery." How many packages were stamped "two-day delivery"?

______ packages were stamped "two-day delivery."

5. Julie sold 423 butterfly stamps and 101 ladybug stamps. How many stamps did Julie sell in all?

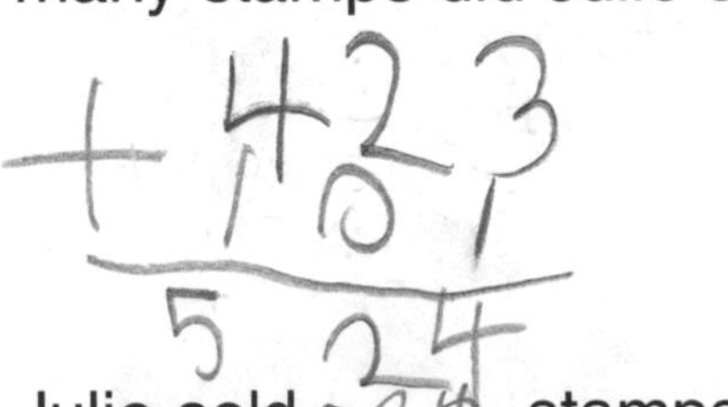

Julie sold ______ stamps in all.

6. Rosa has 200 packages and 357 letters in the post office. How many packages and letters are there in all?

Rosa has ______ packages and letters in all.

Practice 4

Solve each word problem.

1. Cheryl is trying to collect 8,431 pennies. Her parents gave her 1,301 pennies. How many more pennies does Cheryl need to collect to meet her goal?

$$\begin{array}{r} 8{,}431 \\ -\ 1{,}301 \\ \hline \end{array}$$

Cheryl needs to collect _____ more pennies.

2. On Monday, Dwight collected 3,050 pennies. On Tuesday, Dwight collected 3,438 pennies. How many pennies did Dwight collect in all?

$$\begin{array}{r} 3{,}050 \\ +\ 3{,}438 \\ \hline \end{array}$$

Dwight collected _____ pennies in all.

3. Kirby found 5,211 pennies in his piggy bank and 2,506 pennies in his lucky sock. How many pennies did Kirby find in all?

Kirby found _____ pennies in all.

4. Alexandria collected pennies from her neighbors. Mrs. Gibbs gave her 2,102 pennies and Mr. Brown gave her 1,886 pennies. How many pennies did Alexandria collect?

Alexandria collected _____ pennies.

5. Jenny collected 6,533 pennies from the fountain. She collected 3,301 pennies from the top part of the fountain. How many pennies did she collect from the bottom part of the fountain?

She collected _____ pennies from the bottom part of the fountain.

6. Joshua earns a penny for every page he reads. He read 2,110 pages last week and 3,239 pages this week. How many pennies has Joshua earned?

Joshua has earned _____ pennies.

Practice 5

Solve each word problem.

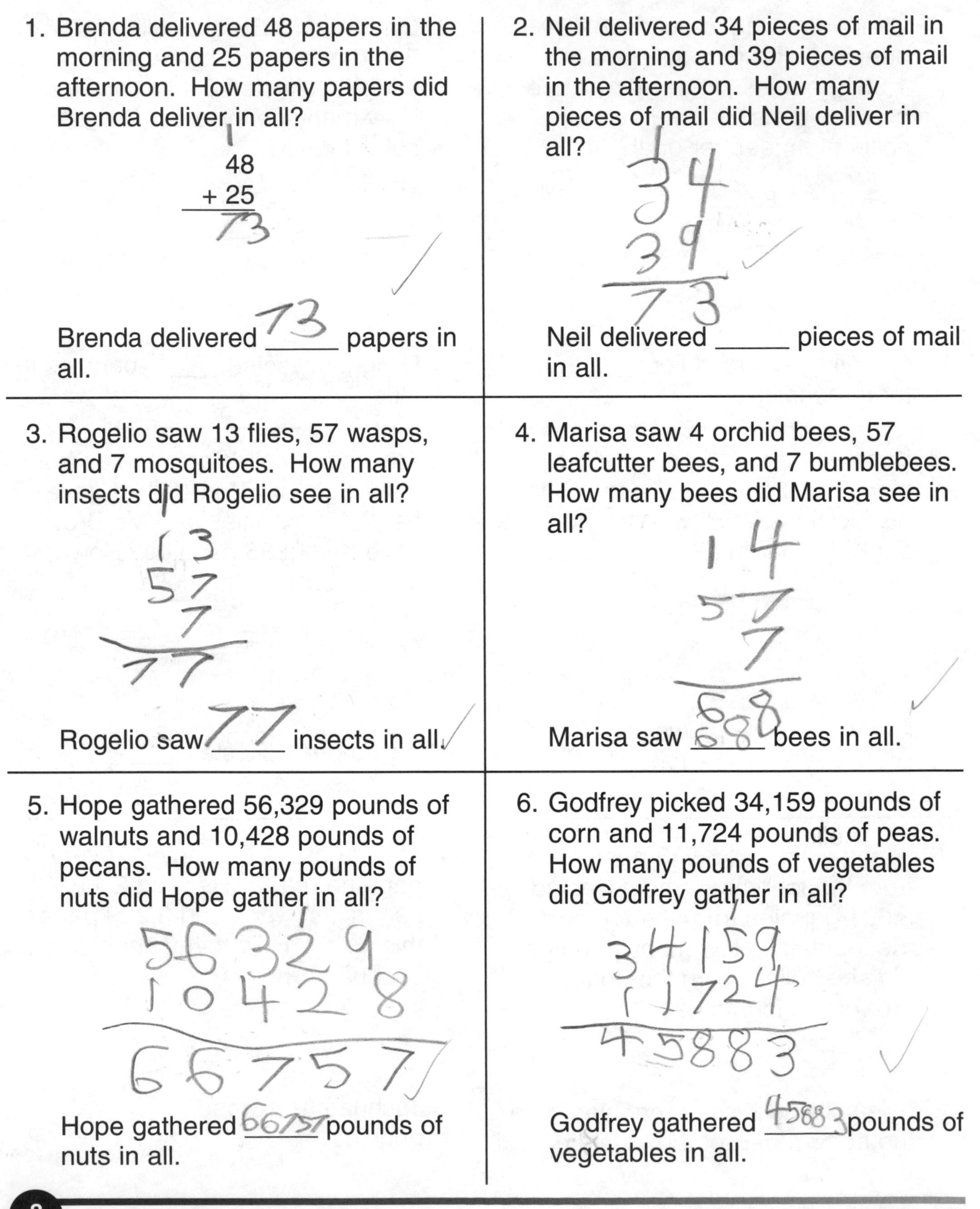

1. Brenda delivered 48 papers in the morning and 25 papers in the afternoon. How many papers did Brenda deliver in all?

$$\begin{array}{r} 48 \\ +\ 25 \\ \hline \end{array}$$

Brenda delivered ______ papers in all.

2. Neil delivered 34 pieces of mail in the morning and 39 pieces of mail in the afternoon. How many pieces of mail did Neil deliver in all?

Neil delivered ______ pieces of mail in all.

3. Rogelio saw 13 flies, 57 wasps, and 7 mosquitoes. How many insects did Rogelio see in all?

Rogelio saw ______ insects in all.

4. Marisa saw 4 orchid bees, 57 leafcutter bees, and 7 bumblebees. How many bees did Marisa see in all?

Marisa saw ______ bees in all.

5. Hope gathered 56,329 pounds of walnuts and 10,428 pounds of pecans. How many pounds of nuts did Hope gather in all?

Hope gathered ______ pounds of nuts in all.

6. Godfrey picked 34,159 pounds of corn and 11,724 pounds of peas. How many pounds of vegetables did Godfrey gather in all?

Godfrey gathered ______ pounds of vegetables in all.

Practice 6

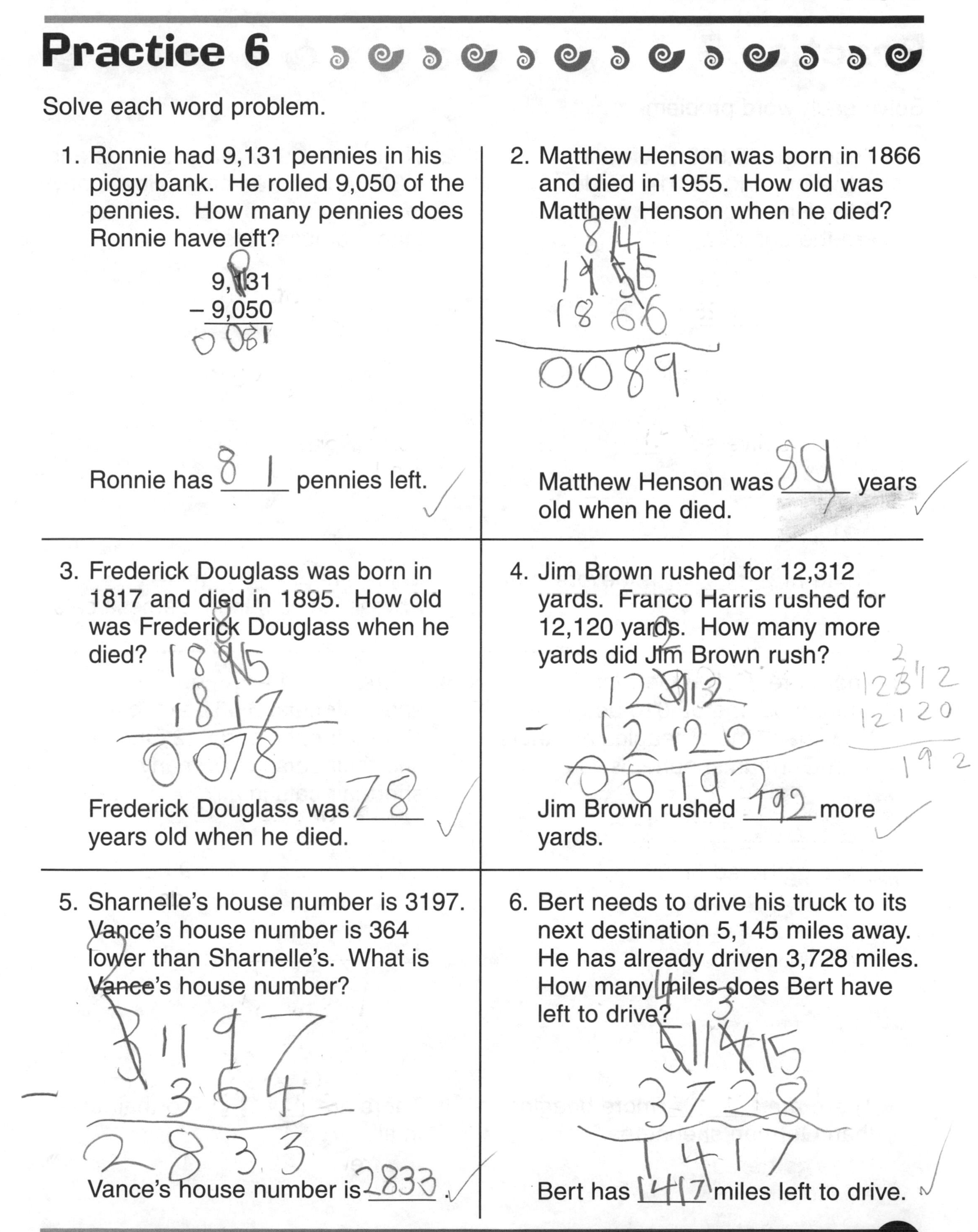

Solve each word problem.

1. Ronnie had 9,131 pennies in his piggy bank. He rolled 9,050 of the pennies. How many pennies does Ronnie have left?

$$\begin{array}{r} 9{,}131 \\ -\ 9{,}050 \\ \hline \end{array}$$

Ronnie has _____ pennies left.

2. Matthew Henson was born in 1866 and died in 1955. How old was Matthew Henson when he died?

Matthew Henson was _____ years old when he died.

3. Frederick Douglass was born in 1817 and died in 1895. How old was Frederick Douglass when he died?

Frederick Douglass was _____ years old when he died.

4. Jim Brown rushed for 12,312 yards. Franco Harris rushed for 12,120 yards. How many more yards did Jim Brown rush?

Jim Brown rushed _____ more yards.

5. Sharnelle's house number is 3197. Vance's house number is 364 lower than Sharnelle's. What is Vance's house number?

Vance's house number is _____.

6. Bert needs to drive his truck to its next destination 5,145 miles away. He has already driven 3,728 miles. How many miles does Bert have left to drive?

Bert has _____ miles left to drive.

Practice 7

Solve each word problem.

1. There are 73,995 Labrador retrievers and 72,914 Golden retrievers. How many retrievers are there in all?

$$\begin{array}{r} 73{,}995 \\ +\ 72{,}914 \\ \hline \end{array}$$

There are ______ retrievers in all.

2. There are 1,979 Siamese cats and 4,349 Abyssinian cats. How many more Abyssinian cats are there than Siamese cats?

$$\begin{array}{r} 4{,}349 \\ -\ 1{,}979 \\ \hline \end{array}$$

There are ______ more Abyssinian cats than Siamese cats.

3. There are 33,108 German shepherds and 52,610 beagles. How many more beagles are there than German shepherds?

There are ______ more beagles than German shepherds.

4. There are 2,112 American shorthair cats, 6,031 exotic shorthair cats, and 942 Oriental shorthair cats. How many shorthair cats in all?

There are ______ shorthair cats in all.

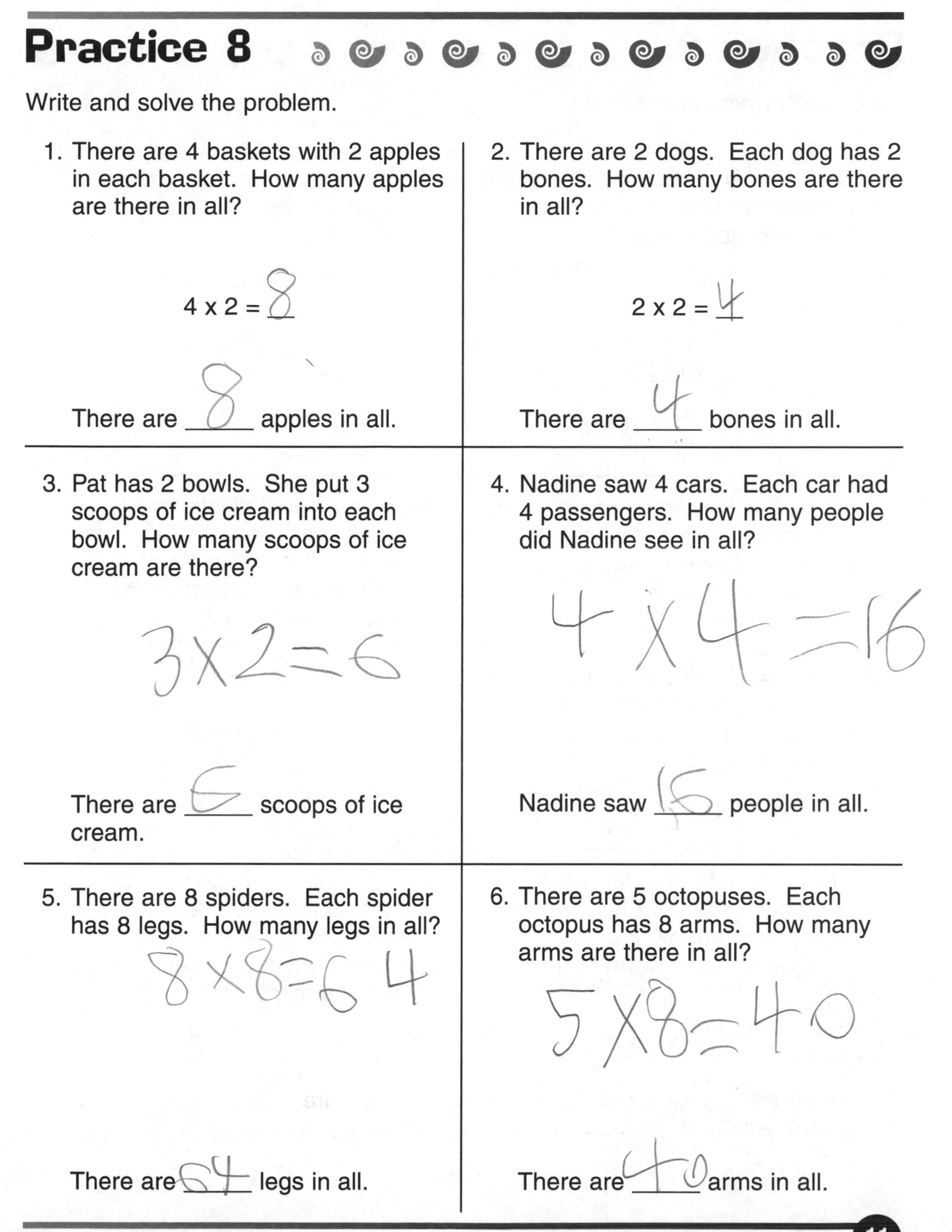

Practice 8

Write and solve the problem.

1. There are 4 baskets with 2 apples in each basket. How many apples are there in all?

   4 x 2 = 8

   There are 8 apples in all.

2. There are 2 dogs. Each dog has 2 bones. How many bones are there in all?

   2 x 2 = 4

   There are 4 bones in all.

3. Pat has 2 bowls. She put 3 scoops of ice cream into each bowl. How many scoops of ice cream are there?

   3 x 2 = 6

   There are 6 scoops of ice cream.

4. Nadine saw 4 cars. Each car had 4 passengers. How many people did Nadine see in all?

   4 x 4 = 16

   Nadine saw 16 people in all.

5. There are 8 spiders. Each spider has 8 legs. How many legs in all?

   8 x 8 = 64

   There are 64 legs in all.

6. There are 5 octopuses. Each octopus has 8 arms. How many arms are there in all?

   5 x 8 = 40

   There are 40 arms in all.

Practice 9

Fill in the correct answer circle.

1. Penny has 5 pounds of newspapers. Each pound is worth a penny. How much money are the newspapers worth?

   - ● 5 x 1¢ = 5¢
   - ○ 5 x 2¢ = 10¢
   - ○ 5 x 5¢ = 25¢

2. The recycling center pays 3¢ for each can. Peter has 8 cans. How much money are the cans worth?

   - ○ 8 x 8¢ = 64¢
   - ○ 8 x 1¢ = 8¢
   - ● 8 x 3¢ = 24¢

3. Nicole has 4 grocery sacks. Inside each grocery sack there are 6 plastic containers. How many containers are there in all?

   - ● 6 x 4 = 24
   - ○ 4 x 4 = 16
   - ○ 6 x 6 = 36

4. Ramon has 3 plastic milk containers. Paulette has 3 times the number of containers than Ramon has. How many containers does Paulette have?

   - ○ 3 x 1 = 3
   - ○ 3 x 2 = 6
   - ● 3 x 3 = 9

5. Coco has 5 cardboard boxes. Each cardboard box is worth 6¢. How much money are the cardboard boxes worth?

   - ○ 5 x 5¢ = 25¢
   - ○ 5 x 6¢ = 30¢
   - ○ 5 x 1¢ = 5¢

6. Todd has 9 magazines. Each magazine can be recycled and made into 3 newspapers. How many newspapers can be made with the magazines?

   - ○ 3 x 3 = 9
   - ● 3 x 9 = 27
   - ○ 7 x 3 = 21

Practice 10

Read each word problem. Write and solve the multiplication problem. Fill in the correct answer circle.

1. There are 4 dozen eggs. How many eggs in all? (*Hint:* There are 12 eggs in a dozen.)

   12 eggs ○ 48 eggs ● 16 eggs ○

2. How many hours in 2 days? (*Hint:* There are 24 hours in 1 day.)

   48 hours ● 24 hours ○ 2 hours ○

3. How many months in 4 years? (*Hint:* There are 12 months in 1 year.)

   48 months ● 84 months ○ 24 months ○

4. How many days in 11 weeks? (*Hint:* There are 7 days in 1 week.)

   18 days ○ 77 days ● 81 days ○

5. How many days in 1 year?

   730 days ○ 365 days ● 12 months ○

6. One month has 30 days. How many days in 7 months?

   37 days ○ 201 days ○ 210 days ●

7. How many inches are in 5 feet? (*Hint:* There are 12 inches in 1 foot.)

   17 inches ○ 60 inches ● 66 inches ○

8. There are 3 feet in 1 yard. How many feet are in 18 yards?

   54 feet ● 45 feet ○ 22 feet ○

Practice 11

Solve each division problem. Fill in the correct answer circle.

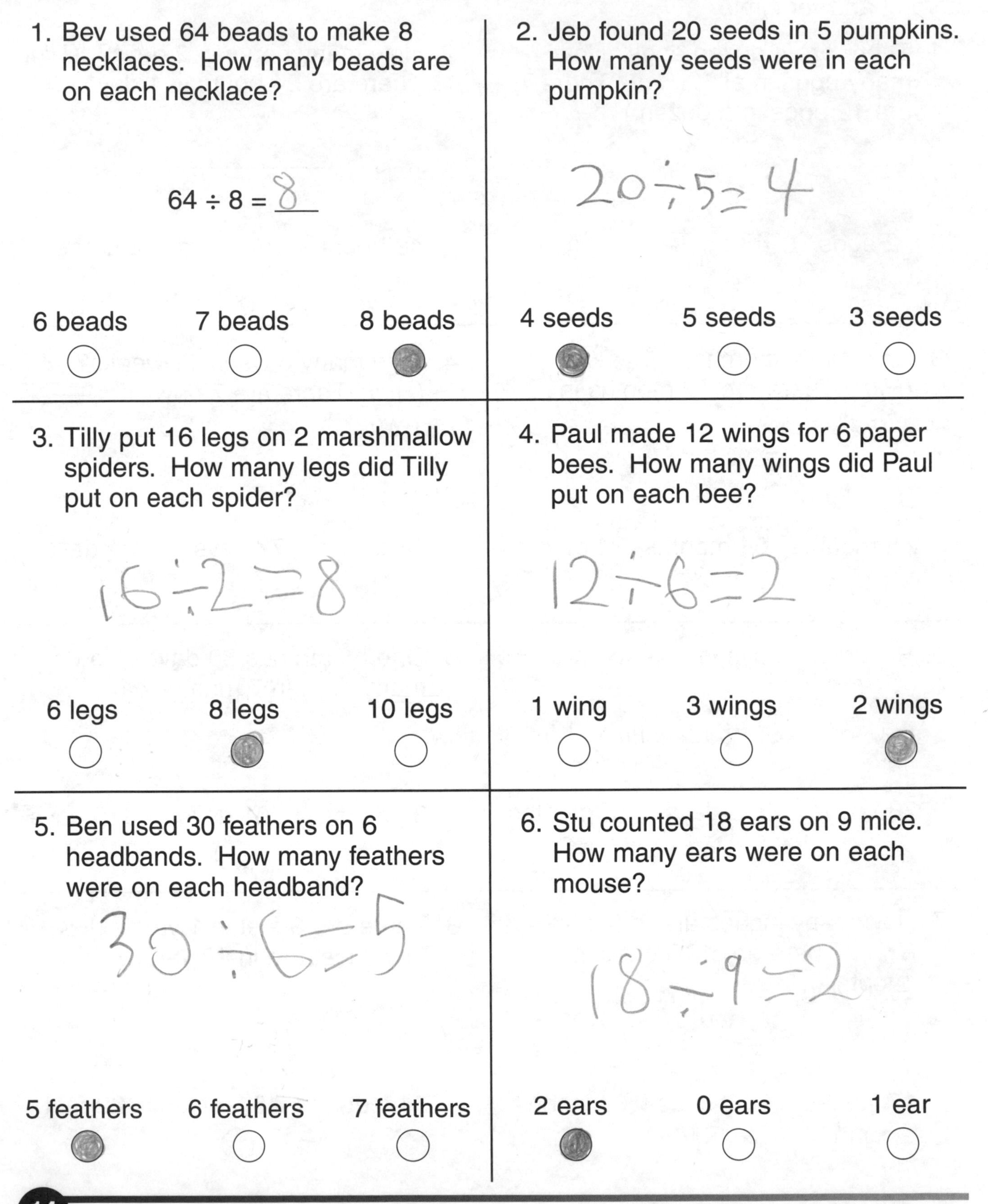

1. Bev used 64 beads to make 8 necklaces. How many beads are on each necklace?

   $64 \div 8 =$ 8

   6 beads ○ 7 beads ○ 8 beads ●

2. Jeb found 20 seeds in 5 pumpkins. How many seeds were in each pumpkin?

   $20 \div 5 = 4$

   4 seeds ● 5 seeds ○ 3 seeds ○

3. Tilly put 16 legs on 2 marshmallow spiders. How many legs did Tilly put on each spider?

   $16 \div 2 = 8$

   6 legs ○ 8 legs ● 10 legs ○

4. Paul made 12 wings for 6 paper bees. How many wings did Paul put on each bee?

   $12 \div 6 = 2$

   1 wing ○ 3 wings ○ 2 wings ●

5. Ben used 30 feathers on 6 headbands. How many feathers were on each headband?

   $30 \div 6 = 5$

   5 feathers ● 6 feathers ○ 7 feathers ○

6. Stu counted 18 ears on 9 mice. How many ears were on each mouse?

   $18 \div 9 = 2$

   2 ears ● 0 ears ○ 1 ear ○

Practice 12

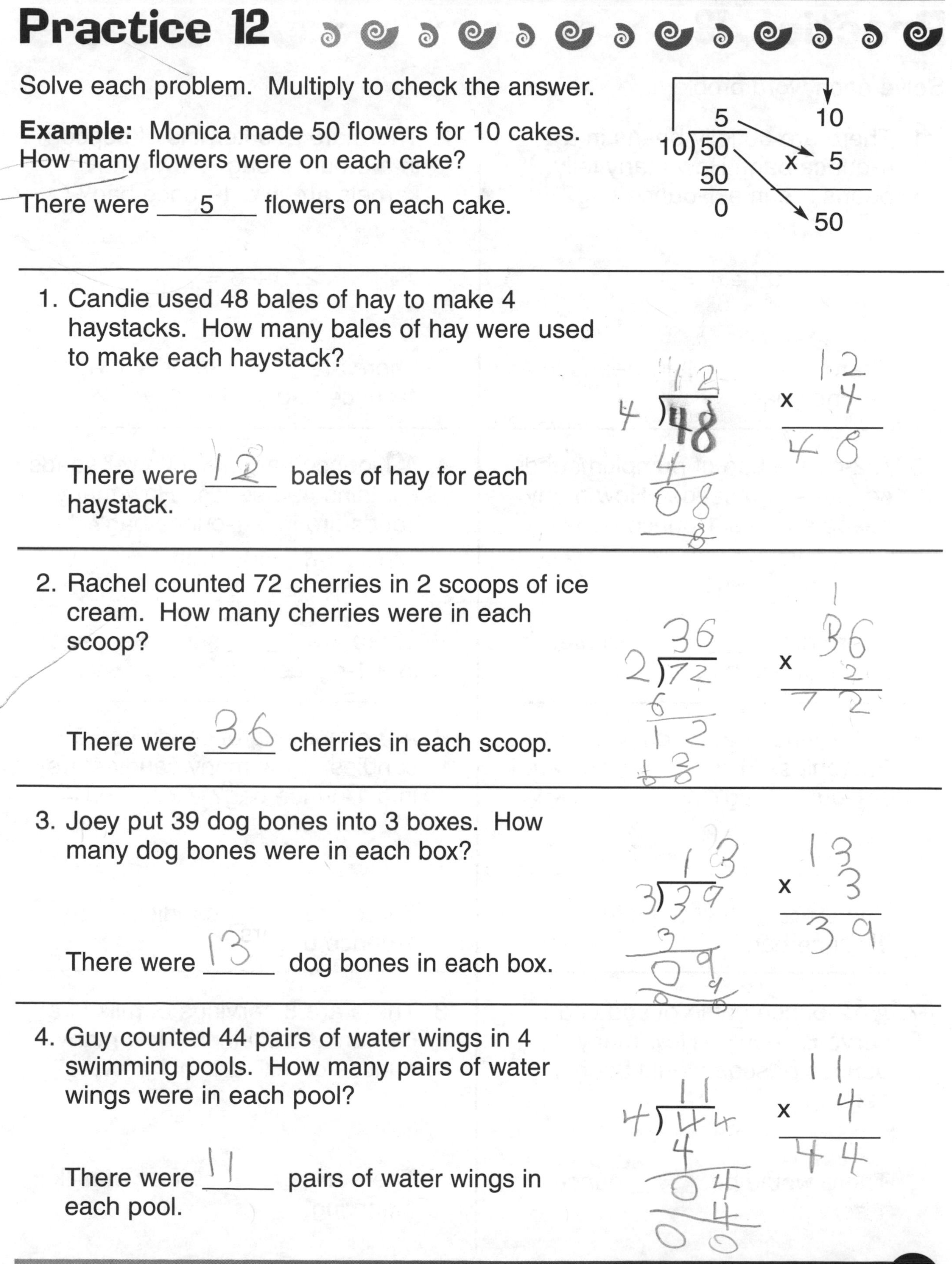

Solve each problem. Multiply to check the answer.

Example: Monica made 50 flowers for 10 cakes. How many flowers were on each cake?

There were ___5___ flowers on each cake.

$$10\overline{)50} = 5 \qquad 10 \times 5 = 50$$

1. Candie used 48 bales of hay to make 4 haystacks. How many bales of hay were used to make each haystack?

   There were ______ bales of hay for each haystack.

2. Rachel counted 72 cherries in 2 scoops of ice cream. How many cherries were in each scoop?

   There were ______ cherries in each scoop.

3. Joey put 39 dog bones into 3 boxes. How many dog bones were in each box?

   There were ______ dog bones in each box.

4. Guy counted 44 pairs of water wings in 4 swimming pools. How many pairs of water wings were in each pool?

   There were ______ pairs of water wings in each pool.

Practice 13

Solve each word problem.

1. There are 60 jelly beans in a 4-ounce bag. How many jelly beans are in a 1-ounce bag?

   $60 \div 4 =$ ______

   There are ______ jelly beans in a 1-ounce bag.

2. There are 270 kernels of popcorn in a 3-ounce bag. How many kernels are in a 1-ounce bag?

   $270 \div 3 =$ ______

   There are ______ kernels in a 1-ounce bag.

3. A 2-ounce bag of pumpkin seeds contains 100 seeds. How many seeds are in a 1-ounce bag?

   There are ______ pumpkin seeds in a 1-ounce bag.

4. A 7-ounce bag of sunflower seeds contains 490 seeds. How many seeds are in a 1-ounce bag?

   There are ______ sunflower seeds in a 1-ounce bag.

5. A 10-ounce bag of chips contains 250 chips. How many chips are in a 1-ounce bag?

   There are ______ chips in a 1-ounce bag.

6. A 9-ounce bag of candy has 99 candies. How many candies are in a 1-ounce bag?

   There are ______ candies in a 1-ounce bag.

7. A 12-ounce bottle of soda can serve 6 people. How many ounces of soda would be in 1 serving?

   There would be ______ ounces in 1 serving.

8. There are 8 servings of milk in a 64-ounce container. How many ounces of milk are in 1 serving?

   There are ______ ounces of milk in 1 serving.

Practice 14

Solve each word problem. Fill in the correct answer circle.

1. The small jar contains 23 buttons. The large jar has 10 times the number of buttons. How many buttons are in the large jar?

   23 x 10 = ____

   23 ○ 203 ○ 230 ○

2. Kevin has 150 tickets. Each ride costs 10 tickets. How many rides can Kevin go on?

   150 ÷ 10 = ____

   10 ○ 15 ○ 51 ○

3. Ashley weighs 80 pounds. That is 10 times more than her baby brother, Matthew. How much does Matthew weigh?

   7 lbs. ○ 8 lbs. ○ 9 lbs. ○

4. A regular light bulb can burn for 40 hours. The new and improved long-lasting light bulb can burn 10 times longer. How long can the light bulb burn?

   40 hours ○ 400 hours ○ 4,000 hours ○

5. There are 10 dimes in one dollar. How many dimes are in ten dollars?

   10 ○ 100 ○ 110 ○

6. Saul has 10 ice cube trays. Each ice cube tray holds 20 ice cubes. How many ice cubes can Saul make?

   100 ○ 200 ○ 300 ○

Practice 15

Solve each word problem. Fill in the answer circle.

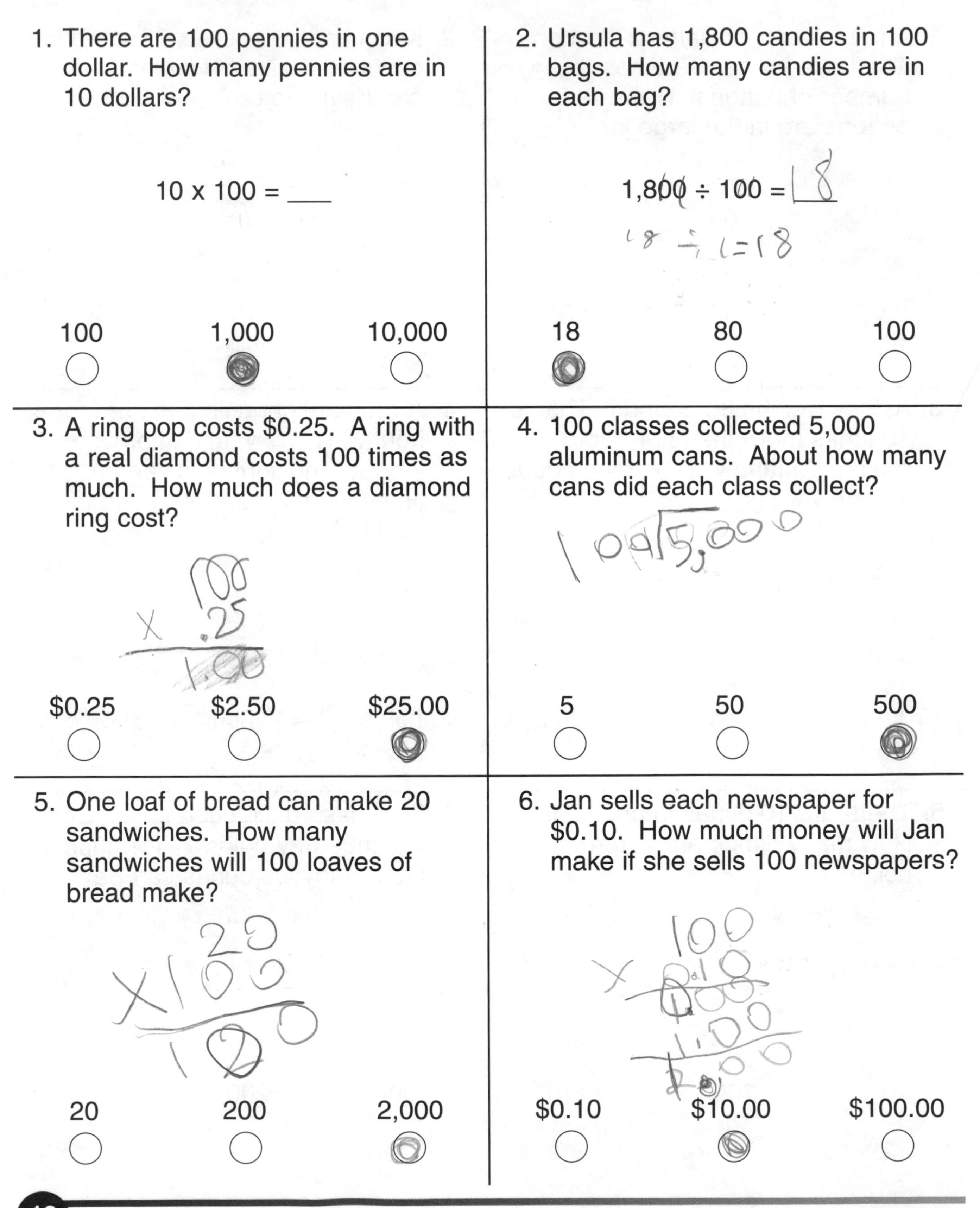

1. There are 100 pennies in one dollar. How many pennies are in 10 dollars?

   10 x 100 = ___

   100 ○ 1,000 ○ 10,000 ○

2. Ursula has 1,800 candies in 100 bags. How many candies are in each bag?

   1,800 ÷ 100 = ___

   18 ○ 80 ○ 100 ○

3. A ring pop costs $0.25. A ring with a real diamond costs 100 times as much. How much does a diamond ring cost?

   $0.25 ○ $2.50 ○ $25.00 ○

4. 100 classes collected 5,000 aluminum cans. About how many cans did each class collect?

   5 ○ 50 ○ 500 ○

5. One loaf of bread can make 20 sandwiches. How many sandwiches will 100 loaves of bread make?

   20 ○ 200 ○ 2,000 ○

6. Jan sells each newspaper for $0.10. How much money will Jan make if she sells 100 newspapers?

   $0.10 ○ $10.00 ○ $100.00 ○

Practice 16

Solve each word problem.

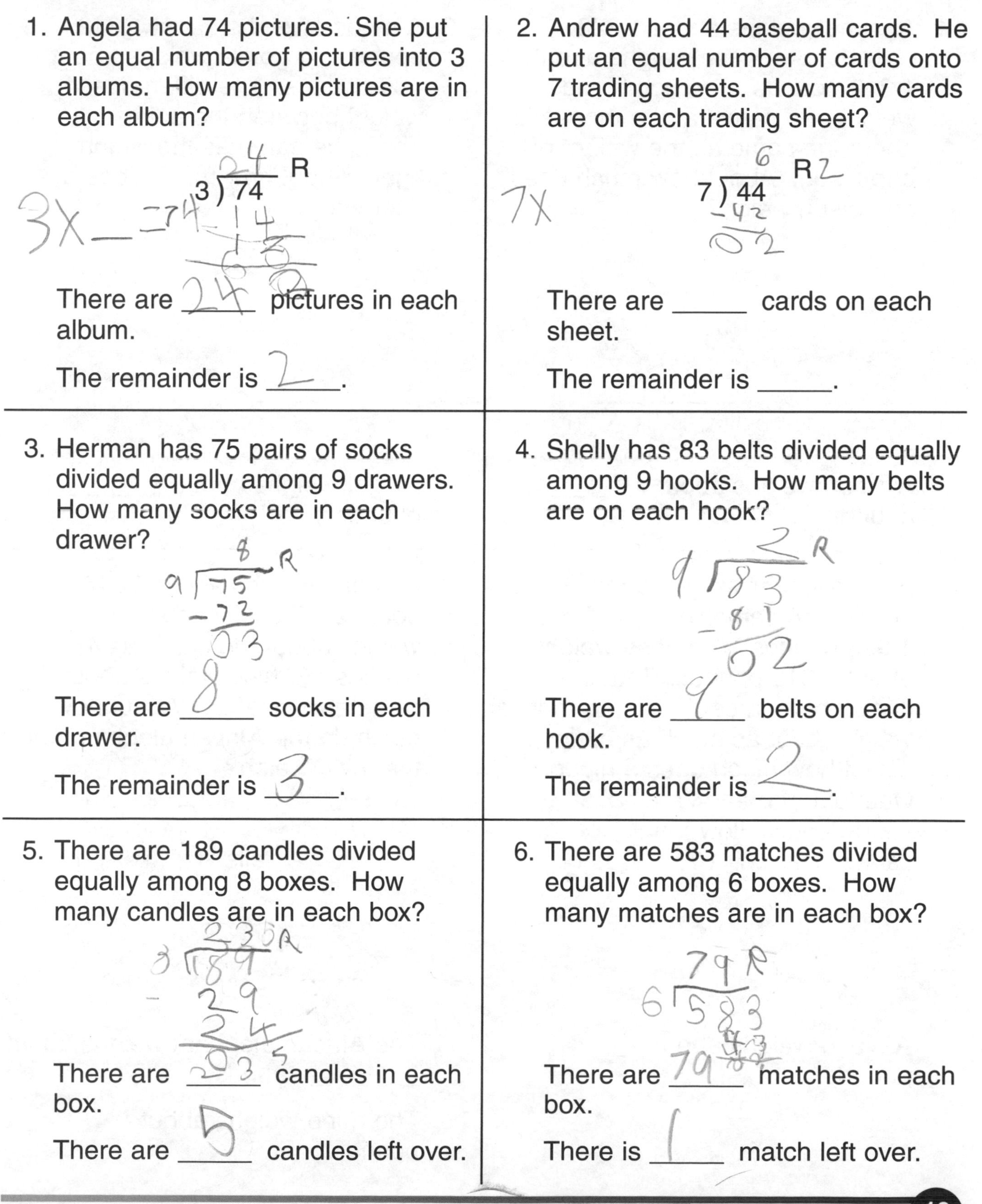

1. Angela had 74 pictures. She put an equal number of pictures into 3 albums. How many pictures are in each album?

   $3\overline{)74}$ R

   There are _____ pictures in each album.

   The remainder is _____.

2. Andrew had 44 baseball cards. He put an equal number of cards onto 7 trading sheets. How many cards are on each trading sheet?

   $7\overline{)44}$ R

   There are _____ cards on each sheet.

   The remainder is _____.

3. Herman has 75 pairs of socks divided equally among 9 drawers. How many socks are in each drawer?

   There are _____ socks in each drawer.

   The remainder is _____.

4. Shelly has 83 belts divided equally among 9 hooks. How many belts are on each hook?

   There are _____ belts on each hook.

   The remainder is _____.

5. There are 189 candles divided equally among 8 boxes. How many candles are in each box?

   There are _____ candles in each box.

   There are _____ candles left over.

6. There are 583 matches divided equally among 6 boxes. How many matches are in each box?

   There are _____ matches in each box.

   There is _____ match left over.

Practice 17

Solve each word problem.

1. A camel weighs about 1,323 pounds. A moose weighs about 1,312 pounds. The combined weight of a camel and a moose is about the same as the weight of one bison. About how much does one bison weigh?

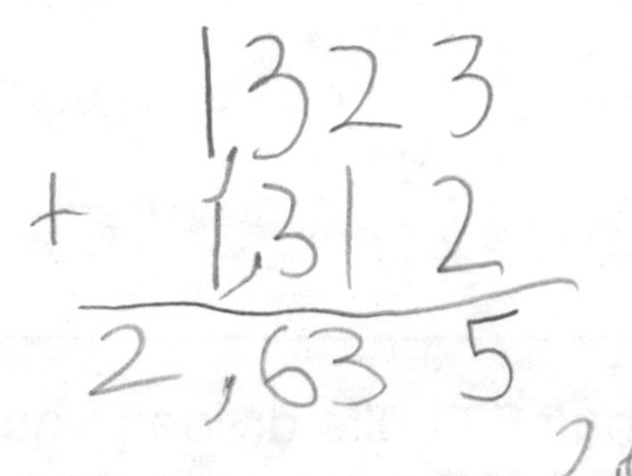

A bison weighs about ________ pounds.

2. A moose weighs about 1,312 pounds. A gorilla weighs about 485 pounds. The difference in weight between the animals is about the same as the weight of a tiger. About how much does a tiger weigh?

A tiger weighs about ________ pounds.

3. A grizzly bear weighs about 1,720 pounds. A camel weighs about 1,323 pounds. A moose weighs about 1,312 pounds. Together, a grizzly bear, a camel, and a moose weigh about as much as a hippo. About how much does a hippo weigh?

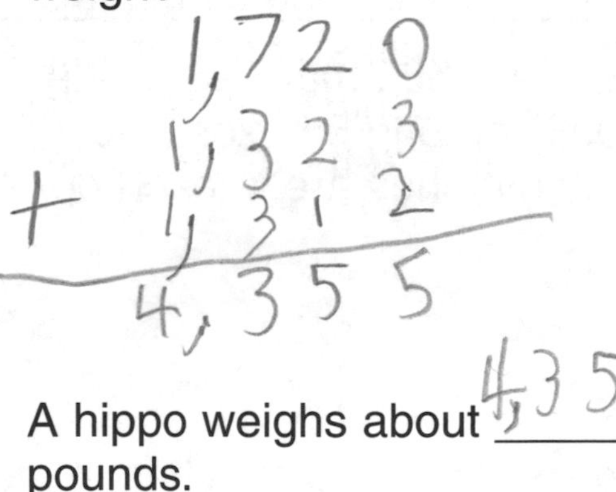

A hippo weighs about ________ pounds.

4. A giraffe weighs about 2,646 pounds. The African elephant weighs about the same as 4 giraffes. A rhino weighs about the same as 3 giraffes. About how much do the African elephant and the rhino weigh?

The African elephant weighs about ________ pounds.

The rhino weighs about ________ pounds.

Practice 18

Solve each word problem. Fill in the correct answer circle.

1. Jeanette spent 3 minutes brushing her teeth, 5 minutes brushing her hair, and a quarter of an hour getting dressed. How many minutes did it take Jeanette to get ready?

   10 min. ○ 23 min. ○ 33 min. ○

2. Noah cooked the meatballs for half an hour and the spaghetti for a quarter of an hour. How many minutes did Noah spend cooking the spaghetti and meatballs?

   15 min. ○ 30 min. ○ 45 min. ○

3. Paige spent a quarter of an hour painting the wall red and 20 minutes painting the ceiling orange. How many minutes did Paige spend painting the wall and the ceiling?

   35 min. ○ 40 min. ○ 45 min. ○

4. Drew spent a quarter of an hour vacuuming the car, a quarter of an hour washing the car, and a quarter of an hour waxing the car. How many minutes did Drew spend on cleaning the car?

   15 min. ○ 30 min. ○ 45 min. ○

5. Stacy spent 10 minutes cleaning her own room, 10 minutes cleaning the house, and 5 minutes raking the leaves. What is the total number of minutes Stacy spent cleaning?

   25 min. ○ 50 min. ○ 75 min. ○

6. Andre spent 8 minutes getting dressed, a half an hour eating, and 1 minute brushing his hair. How long did it take for Andre to get ready?

   24 min. ○ 20 min. ○ 39 min. ○

Practice 19

Solve each word problem.

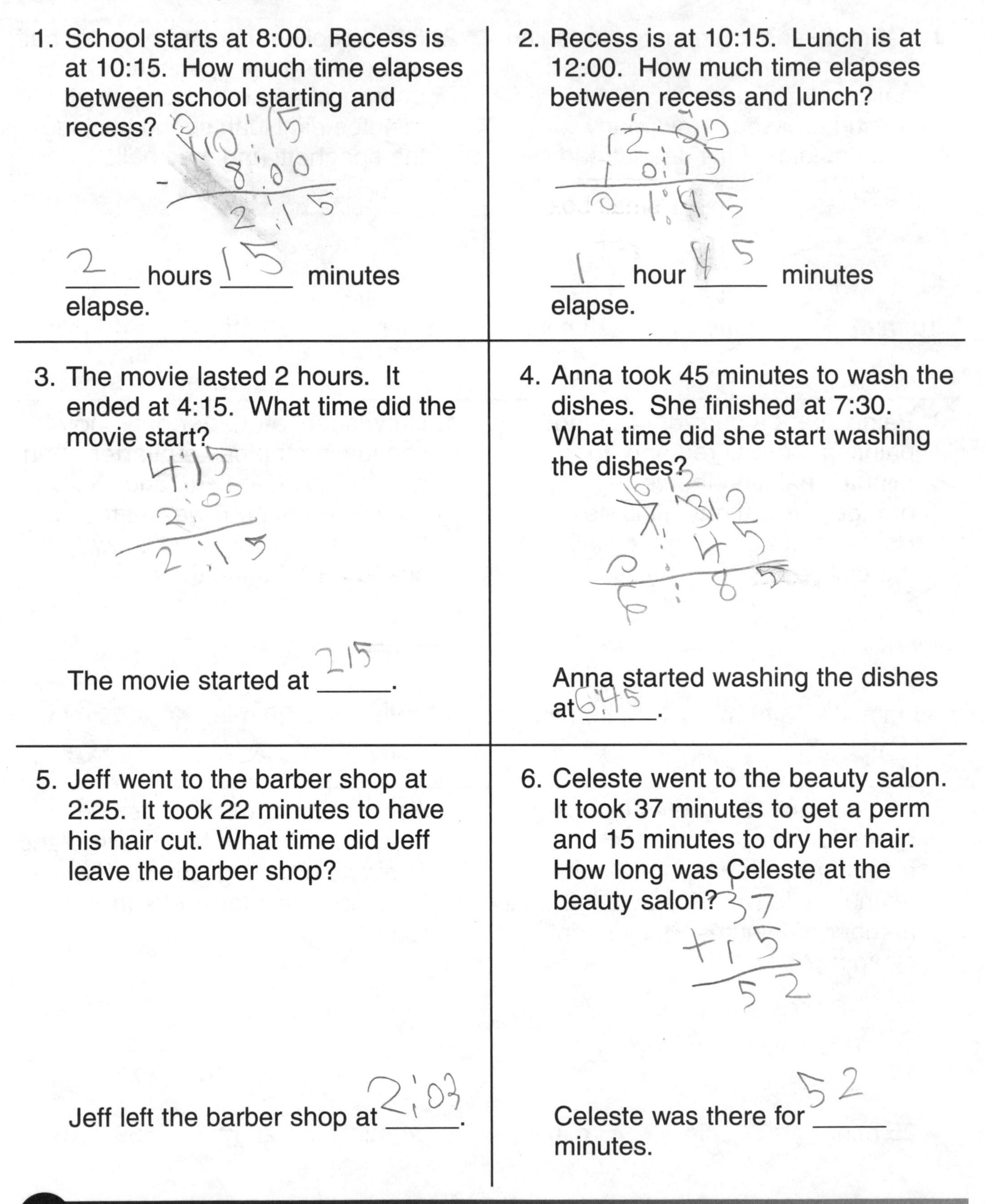

1. School starts at 8:00. Recess is at 10:15. How much time elapses between school starting and recess?

   _____ hours _____ minutes elapse.

2. Recess is at 10:15. Lunch is at 12:00. How much time elapses between recess and lunch?

   _____ hour _____ minutes elapse.

3. The movie lasted 2 hours. It ended at 4:15. What time did the movie start?

   The movie started at _____.

4. Anna took 45 minutes to wash the dishes. She finished at 7:30. What time did she start washing the dishes?

   Anna started washing the dishes at _____.

5. Jeff went to the barber shop at 2:25. It took 22 minutes to have his hair cut. What time did Jeff leave the barber shop?

   Jeff left the barber shop at _____.

6. Celeste went to the beauty salon. It took 37 minutes to get a perm and 15 minutes to dry her hair. How long was Celeste at the beauty salon?

   Celeste was there for _____ minutes.

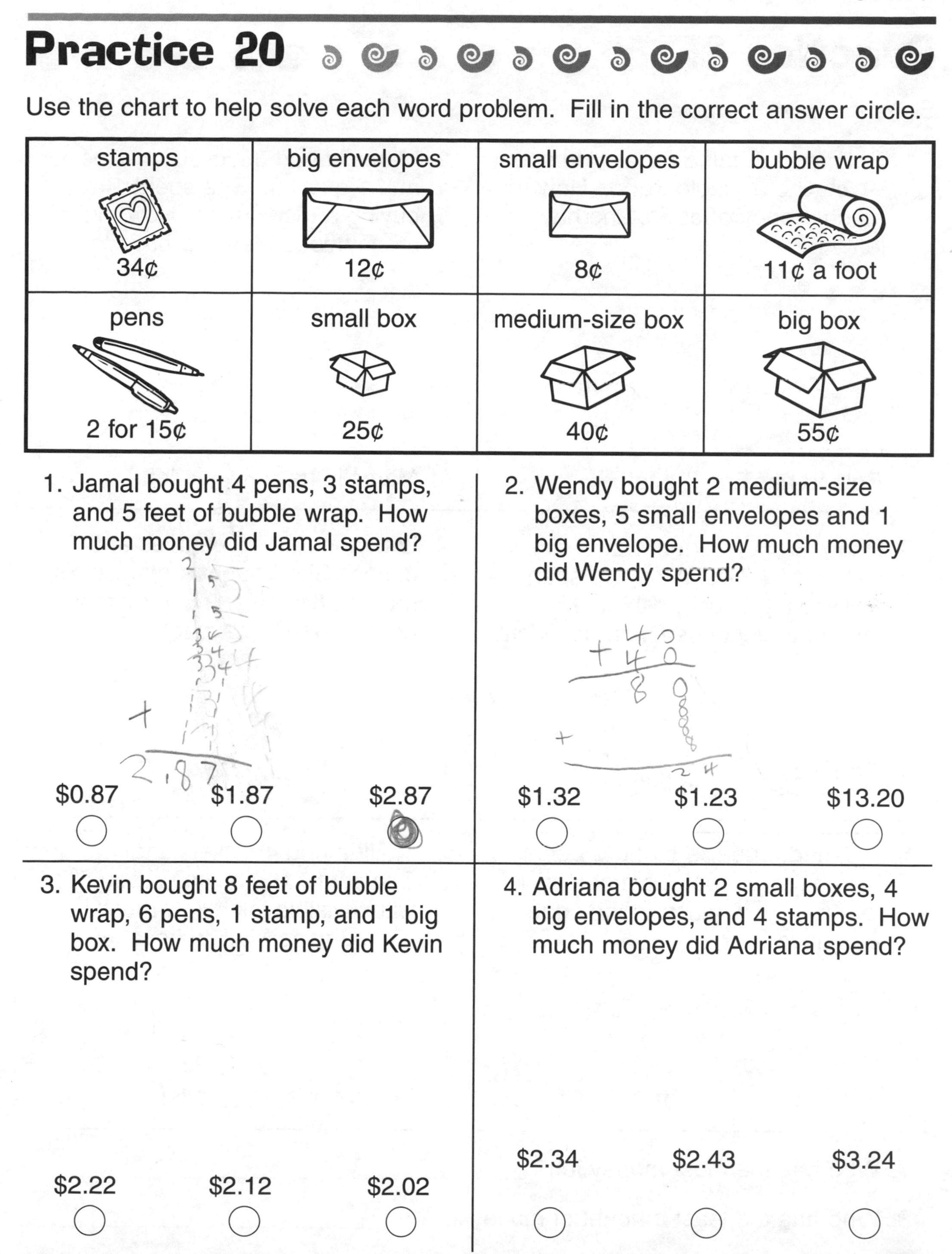

Practice 20

Use the chart to help solve each word problem. Fill in the correct answer circle.

stamps	big envelopes	small envelopes	bubble wrap
34¢	12¢	8¢	11¢ a foot
pens	**small box**	**medium-size box**	**big box**
2 for 15¢	25¢	40¢	55¢

1. Jamal bought 4 pens, 3 stamps, and 5 feet of bubble wrap. How much money did Jamal spend?

   ○ $0.87 ○ $1.87 ○ $2.87

2. Wendy bought 2 medium-size boxes, 5 small envelopes and 1 big envelope. How much money did Wendy spend?

   ○ $1.32 ○ $1.23 ○ $13.20

3. Kevin bought 8 feet of bubble wrap, 6 pens, 1 stamp, and 1 big box. How much money did Kevin spend?

   ○ $2.22 ○ $2.12 ○ $2.02

4. Adriana bought 2 small boxes, 4 big envelopes, and 4 stamps. How much money did Adriana spend?

   ○ $2.34 ○ $2.43 ○ $3.24

Practice 21

Solve each word problem.

1. Roland had half a dollar. He spent 9¢ buying a magic wand. How much money does Roland have left?

   $$\begin{array}{r} 50¢ \\ -\ 9¢ \\ \hline \end{array}$$

   Roland has _____¢ left.

2. Marcella had 3 nickels, 1 dime, and 1 quarter. She spent 14¢ buying a trick flower. How much money does Marcella have left?

   Marcella has _____¢ left.

3. Dean had 1 quarter, 4 dimes, 1 nickel, and 8 pennies. He spent 29¢ buying a lucky rabbit. How much money does Dean have left?

   Dean has _____¢ left.

4. Louanne had 15 pennies and 7 dimes. She bought a magician's hat for 28¢. How much money does Louanne have left?

   Louanne has _____¢ left.

5. R.J. had 3 quarters and 3 pennies. He bought a book on magic tricks for 49¢. How much money does R.J. have left?

   R.J. has _____¢ left.

6. Matilda had 4 nickels and 1 penny. She spent 18¢ buying a pen filled with invisible ink. How much money does Matilda have left?

   Matilda has _____¢ left.

7. Who has the most money left? ______________________________

8. Who has the least amount of money left? ______________________________

Practice 22

Use the chart to help solve each word problem. Fill in the correct answer circle.

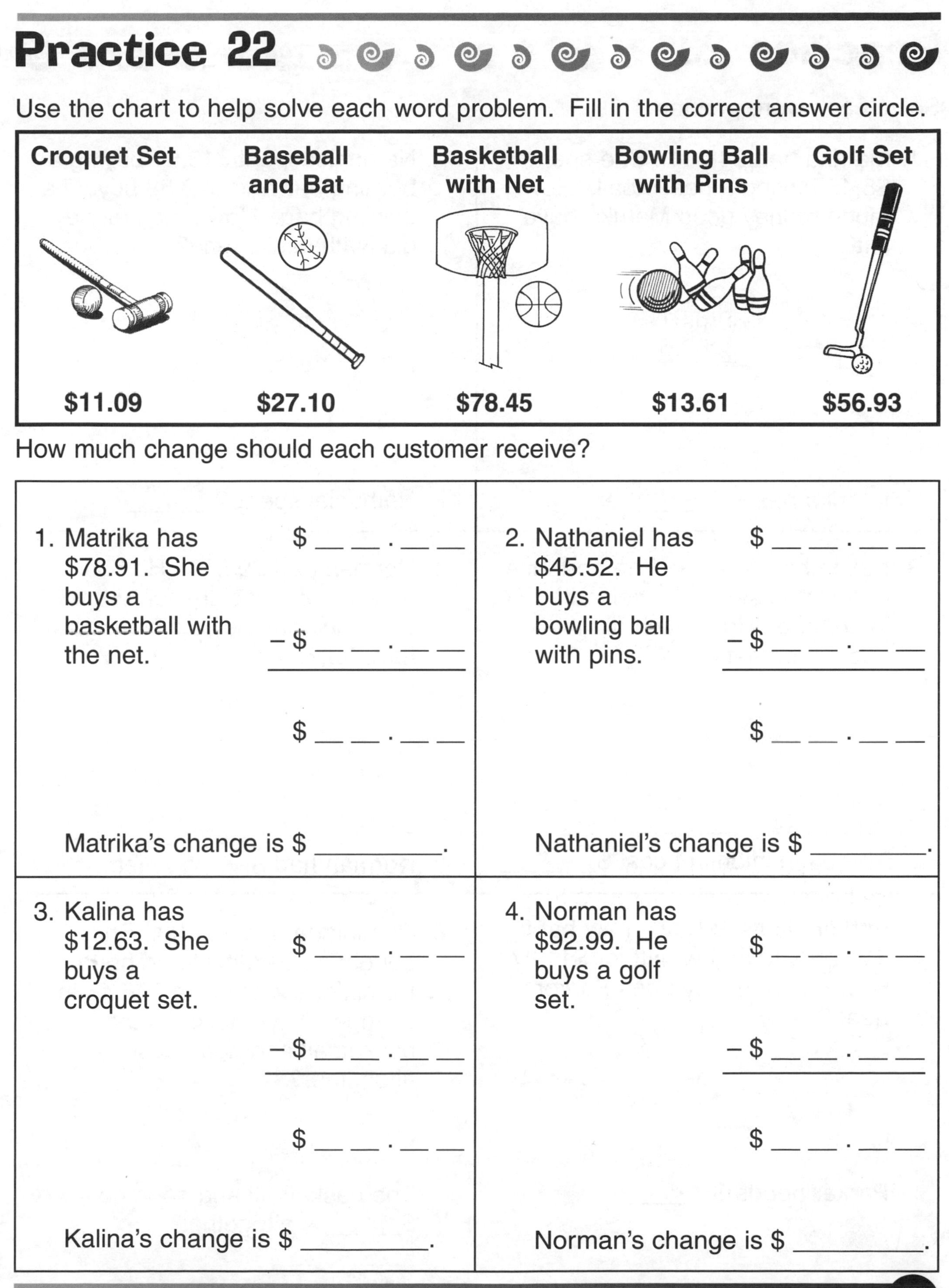

Croquet Set	Baseball and Bat	Basketball with Net	Bowling Ball with Pins	Golf Set
$11.09	$27.10	$78.45	$13.61	$56.93

How much change should each customer receive?

1. Matrika has $78.91. She buys a basketball with the net.

   \$ __ __ . __ __

   – \$ __ __ . __ __

   \$ __ __ . __ __

   Matrika's change is \$ __________.

2. Nathaniel has $45.52. He buys a bowling ball with pins.

   \$ __ __ . __ __

   – \$ __ __ . __ __

   \$ __ __ . __ __

   Nathaniel's change is \$ __________.

3. Kalina has $12.63. She buys a croquet set.

   \$ __ __ . __ __

   – \$ __ __ . __ __

   \$ __ __ . __ __

   Kalina's change is \$ __________.

4. Norman has $92.99. He buys a golf set.

   \$ __ __ . __ __

   – \$ __ __ . __ __

   \$ __ __ . __ __

   Norman's change is \$ __________.

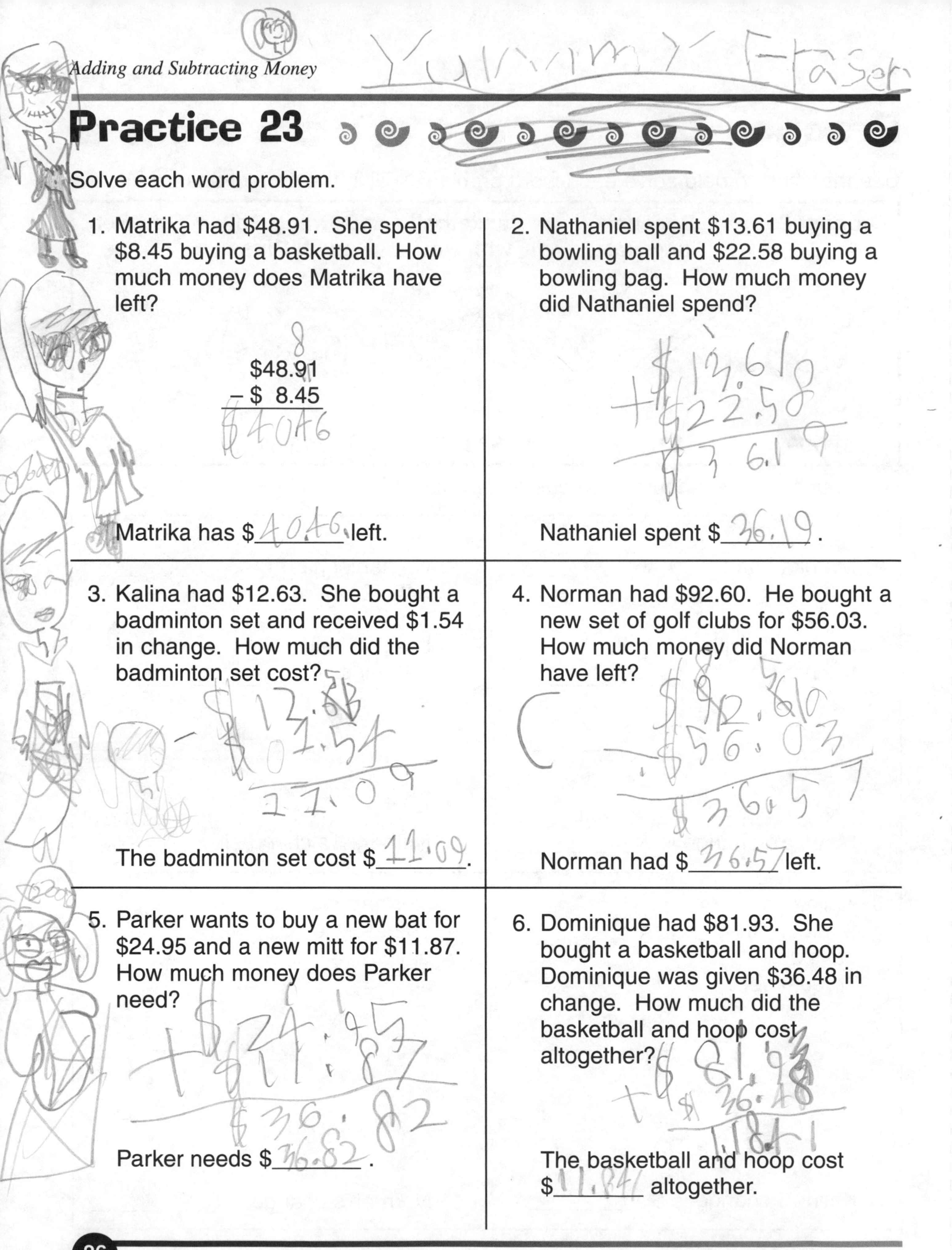

Practice 23

Solve each word problem.

1. Matrika had $48.91. She spent $8.45 buying a basketball. How much money does Matrika have left?

   $48.91
   – $ 8.45

   Matrika has $________ left.

2. Nathaniel spent $13.61 buying a bowling ball and $22.58 buying a bowling bag. How much money did Nathaniel spend?

   Nathaniel spent $________.

3. Kalina had $12.63. She bought a badminton set and received $1.54 in change. How much did the badminton set cost?

   The badminton set cost $________.

4. Norman had $92.60. He bought a new set of golf clubs for $56.03. How much money did Norman have left?

   Norman had $________ left.

5. Parker wants to buy a new bat for $24.95 and a new mitt for $11.87. How much money does Parker need?

   Parker needs $________.

6. Dominique had $81.93. She bought a basketball and hoop. Dominique was given $36.48 in change. How much did the basketball and hoop cost altogether?

   The basketball and hoop cost $________ altogether.

Practice 24

Solve each word problem.

1. Kay bought 1 binder for $1.49. Kay also bought 4 erasers. The erasers sold for 2 for $0.39. How much money did Kay spend?

   $1.49
   $0.39
   + $0.39

   Kay spent _______ .

2. Ray bought 1 calculator for $1.99, a notebook for $0.35, and 3 pencils. The pencils cost $0.10 each. How much money did Ray spend?

   Ray spent _______ .

3. Jay had $1.50. He bought a pencil sharpener for $0.22 and 2 reams of paper. Each ream of paper costs $0.13. How much money did Jay have left?

   Jay had _______ left.

4. Fay had $2.00. She spent $0.15 on a notebook cover and $0.28 for a box of crayons. How much money did Fay have left?

   Fay had _______ left.

5. May bought 6 bookmarks. The bookmarks cost $0.25 for 3. May also bought a folder for $0.11. How much money did May spend?

   May spent _______ .

6. Clay bought a ruler for $0.33 and a pen for $0.10. Clay paid for the items with a one dollar bill. How much change was Clay given?

   Clay was given _______ in change.

7. Gray had $0.71. He spent $0.11 buying a notepad and $0.19 buying a small stapler. How much money did Gray have left?

   Gray had _______ left.

8. Bay bought a pencil box for $0.84, a jumbo pencil for $0.25, and a set of markers for $0.39. How much money did Bay spend?

   Bay spent _______ .

Practice 25

Use the pictures to help you solve the problems.

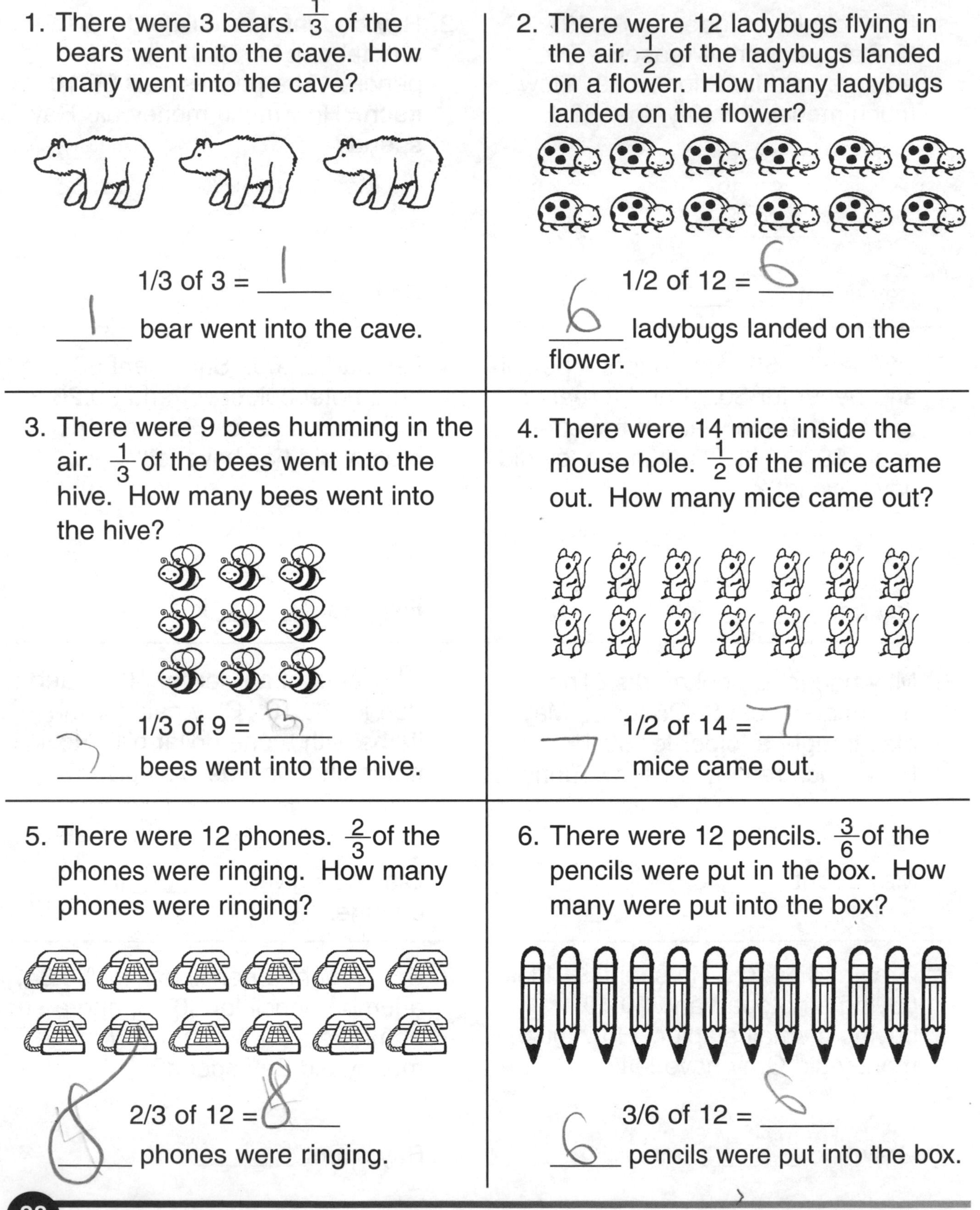

1. There were 3 bears. $\frac{1}{3}$ of the bears went into the cave. How many went into the cave?

   1/3 of 3 = __1__

   __1__ bear went into the cave.

2. There were 12 ladybugs flying in the air. $\frac{1}{2}$ of the ladybugs landed on a flower. How many ladybugs landed on the flower?

   1/2 of 12 = __6__

   __6__ ladybugs landed on the flower.

3. There were 9 bees humming in the air. $\frac{1}{3}$ of the bees went into the hive. How many bees went into the hive?

   1/3 of 9 = __3__

   __3__ bees went into the hive.

4. There were 14 mice inside the mouse hole. $\frac{1}{2}$ of the mice came out. How many mice came out?

   1/2 of 14 = __7__

   __7__ mice came out.

5. There were 12 phones. $\frac{2}{3}$ of the phones were ringing. How many phones were ringing?

   2/3 of 12 = __8__

   __8__ phones were ringing.

6. There were 12 pencils. $\frac{3}{6}$ of the pencils were put in the box. How many were put into the box?

   3/6 of 12 = __6__

   __6__ pencils were put into the box.

Practice 26

Solve each word problem. Shade the parts of the circle to help solve the problem.

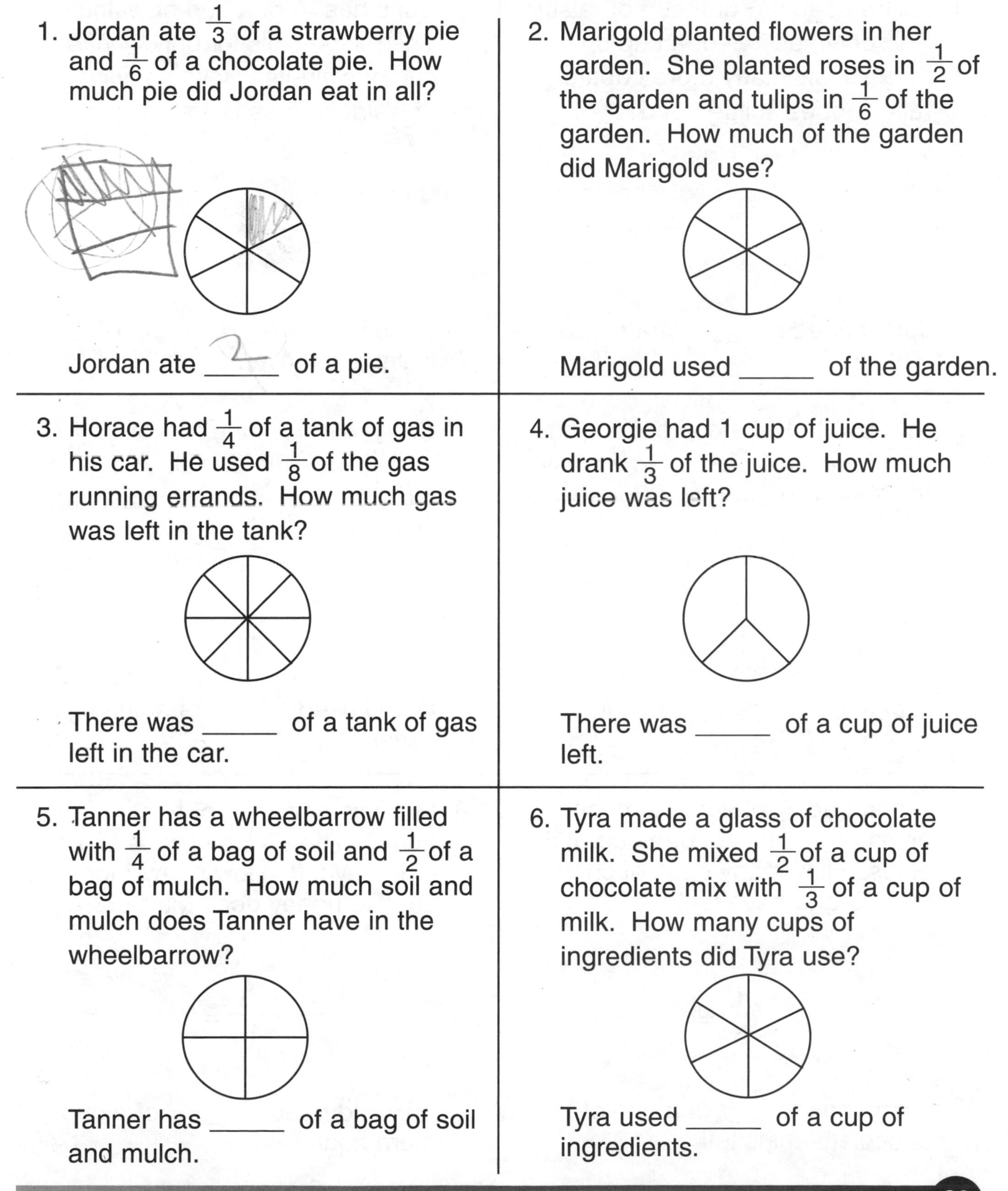

1. Jordan ate $\frac{1}{3}$ of a strawberry pie and $\frac{1}{6}$ of a chocolate pie. How much pie did Jordan eat in all?

   Jordan ate _____ of a pie.

2. Marigold planted flowers in her garden. She planted roses in $\frac{1}{2}$ of the garden and tulips in $\frac{1}{6}$ of the garden. How much of the garden did Marigold use?

   Marigold used _____ of the garden.

3. Horace had $\frac{1}{4}$ of a tank of gas in his car. He used $\frac{1}{8}$ of the gas running errands. How much gas was left in the tank?

   There was _____ of a tank of gas left in the car.

4. Georgie had 1 cup of juice. He drank $\frac{1}{3}$ of the juice. How much juice was left?

   There was _____ of a cup of juice left.

5. Tanner has a wheelbarrow filled with $\frac{1}{4}$ of a bag of soil and $\frac{1}{2}$ of a bag of mulch. How much soil and mulch does Tanner have in the wheelbarrow?

   Tanner has _____ of a bag of soil and mulch.

6. Tyra made a glass of chocolate milk. She mixed $\frac{1}{2}$ of a cup of chocolate mix with $\frac{1}{3}$ of a cup of milk. How many cups of ingredients did Tyra use?

   Tyra used _____ of a cup of ingredients.

Practice 27

Solve each word problem.

1. Joline needs $\frac{8}{9}$ of a cup of raisins. She only has $\frac{4}{9}$ of a cup on hand. How many more cups of raisins does Joline need?

   $$\frac{8}{9} - \frac{4}{9} =$$

   Joline needs _____ of a cup of raisins.

2. Boris has $\frac{7}{8}$ of a cup of walnuts. The recipe calls for only $\frac{3}{8}$ of a cup of walnuts. How many cups of walnuts does Boris have left?

   $$\frac{7}{8} - \frac{3}{8} =$$

   Boris has _____ of a cup of walnuts left over.

3. Kuri used $\frac{3}{7}$ of a cup of flour and $\frac{2}{7}$ of a cup of water. How many cups of ingredients did Kuri use in all?

   $$\frac{3}{7} + \frac{2}{7} =$$

   Kuri used _____ of a cup in all.

4. Darryl used $\frac{8}{11}$ of a cup of sugar and $\frac{2}{11}$ of a cup of milk. How many more cups of sugar than milk did Darryl use?

   $$\frac{8}{11} - \frac{2}{11} =$$

   Darryl used _____ of a cup more sugar.

5. Diana had $\frac{4}{5}$ of a cup of chocolate chips. She ate $\frac{1}{5}$ of a cup of the chips. How many cups of chips does Diana have left?

   $$\frac{4}{5} - \frac{1}{5} =$$

   Diana has _____ of a cup of chocolate chips left.

6. Mayfield needs $\frac{7}{10}$ of a cup of honey. There was only $\frac{3}{10}$ of a cup of honey left. How many more cups of honey does Mayfield need?

   $$\frac{7}{10} - \frac{3}{10} =$$

   Mayfield needs _____ of a cup more honey.

Practice 28

Write the problem and solve it.

1. Clarice ate $\frac{1}{4}$ of a pumpkin pie, and Clarence ate $\frac{2}{4}$ of a pumpkin pie. How much pie did they eat in all?

   They ate _____ of a pie.

2. Pamela and Harrison each ate $\frac{2}{5}$ of a candy bar. How much of the candy bar did they eat in all?

   They ate _____ of a candy bar.

3. The sugar bowl holds $\frac{3}{4}$ of a cup. Marnie used $\frac{1}{4}$ of a cup of sugar. How much sugar is left in the bowl?

   There is _____ cup of sugar left.

4. The container held $\frac{5}{6}$ of a cup of car wax. Harvey used $\frac{3}{6}$ of a cup to wax the car. How much wax is left?

   There is _____ cup of wax left.

5. Ben was given $\frac{5}{8}$ of a pie. He ate $\frac{2}{8}$ of the pie. How much pie was left?

   There was _____ pie left.

6. Julia had $\frac{18}{24}$ of a box of crayons. Her dog ate $\frac{6}{24}$ of the crayons. How many crayons does Julia have left?

   Julia has _____ of a box of crayons left.

Practice 29

Multiply each fraction by the other fraction's denominator. Then add or subtract.

Example $\frac{1}{4} + \frac{2}{3} =$ ______

$$\left(\frac{1}{4} \times \frac{3}{3}\right) + \left(\frac{2}{3} \times \frac{4}{4}\right) = \frac{3}{12} + \frac{8}{12} = \frac{11}{12}$$

1. Bill ate $\frac{1}{4}$ of the turkey. Will ate $\frac{1}{3}$ of the turkey. How much turkey did they eat in all?

   They ate _____ of the turkey.

2. Stephanie ate $\frac{3}{5}$ of the pizza. Jane ate $\frac{1}{6}$ of the pizza. How much more of the pizza did Stephanie eat than Jane?

   Stephanie ate _____ more of the pizza than Jane.

3. Curt ran $\frac{3}{10}$ of a mile and walked $\frac{2}{5}$ of a mile. How far did Curt run and walk?

   Curt ran and walked _____ of a mile.

4. Jack cut off $\frac{1}{5}$ of a board. Marnie cut off $\frac{1}{3}$ of the board. How much of the board did they cut off altogether?

   They cut off _____ of the board.

5. Jennifer ate $\frac{2}{3}$ of the brownies and Emily ate $\frac{3}{5}$ of the brownies. How much did they eat altogether?

   They ate _____ of the brownies altogether.

6. Ron jogged $\frac{3}{4}$ of a mile. James jogged $\frac{4}{5}$ of a mile. How much more of a mile did James jog than Ron?

   James jogged _____ of a mile more.

Practice 30

Read each word problem. How likely is it that the event will take place? Circle *likely* or *unlikely*.

1. Every January 1st, for the last 100 years, Smith's Department Store has held a big, blowout sale. What is the chance that this January 1st, Smith's Department Store will hold the sale?

   Likely Unlikely

2. The Aces have lost the last 10 out of 10 ball games. What is the team's chance of winning the next game?

   Likely Unlikely

3. Damian earns 100% on each week's spelling test. What is Damian's chance of earning 100% on this week's spelling test?

   Likely Unlikely

4. The town of Dry-n-Dusty has not seen a drop of rain in 50 years. What is the chance of rain this week in Dry-n-Dusty?

   Likely Unlikely

5. Every 5 years, millions of ladybugs invade Dusty Gulch. The last invasion was 2 years ago. What is the chance that Dusty Gulch will be invaded by ladybugs this year?

   Likely Unlikely

6. Mary's mom makes her a peanut butter and jelly sandwich for lunch every day. What is Mary's chance of having a peanut butter and jelly sandwich for lunch today?

   Likely Unlikely

Practice 31

Solve each word problem. Fill in the correct answer circle.

1. Elsa typed 200 words in 4 minutes. About how many words can Elsa type in 1 minute?

> 50	About 50	< 50
○	○	○

2. Pierre typed 200 words—but every 5th word was wrong! About how many mistakes did Pierre make?

> 35	About 35	< 35
○	○	○

3. Betty can type 38 words per minute on a typewriter. She can type 3 times as many words on a computer. About how many words can Betty type on a computer?

> 100	About 100	< 100
○	○	○

4. Edgar typed the word "math" 29 times. How many letters did he type?

> 110	About 110	< 110
○	○	○

5. Twelve sentences will fit on one piece of paper. If Eva needs to type 192 sentences, about how many pieces of paper will she need?

> 18	About 18	< 18
○	○	○

6. Brandon can type 18 words in one minute. How many words can Brandon type in 8 minutes?

> 80	About 80	< 80
○	○	○

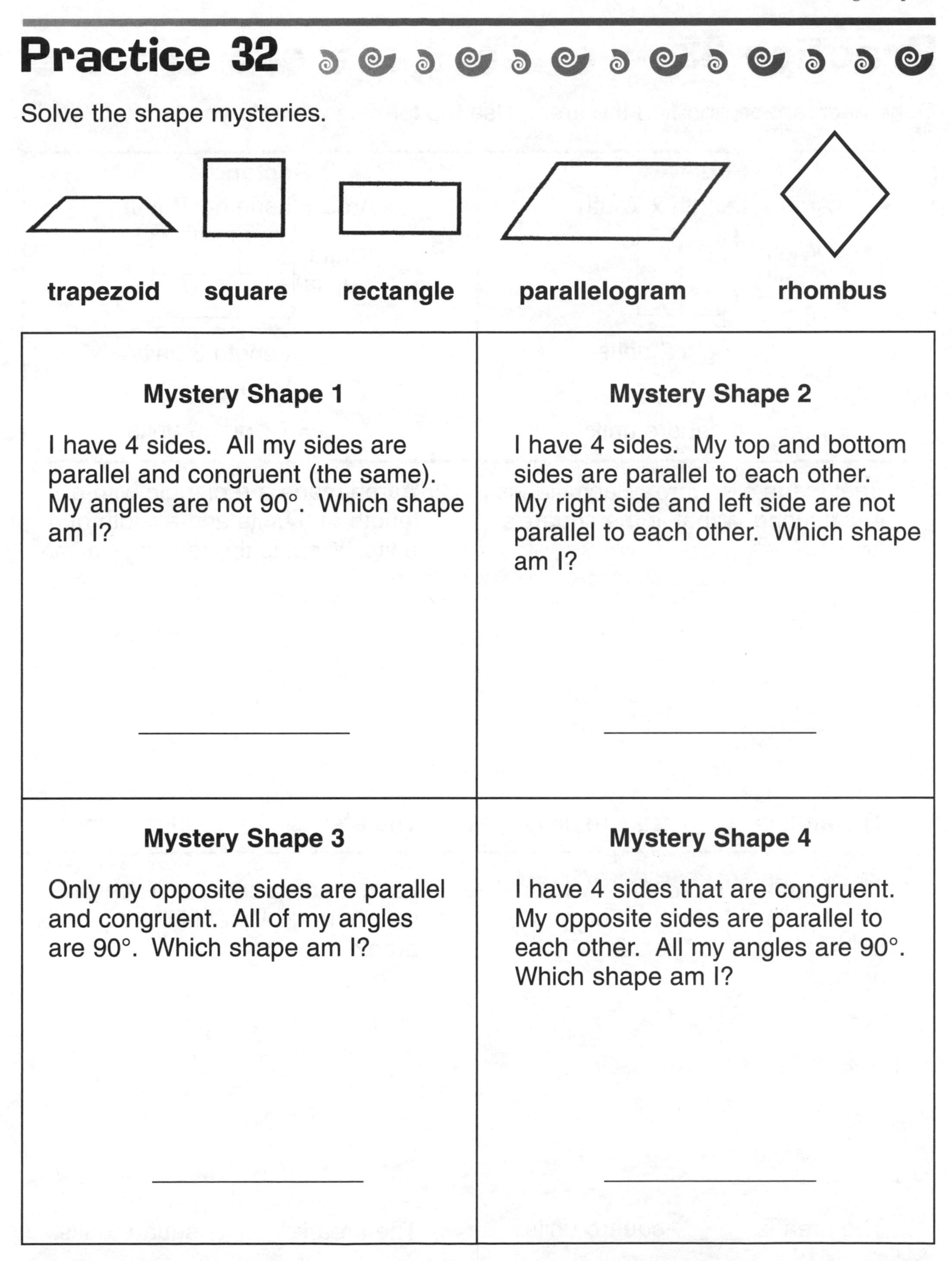

Practice 32

Solve the shape mysteries.

Mystery Shape 1	**Mystery Shape 2**
I have 4 sides. All my sides are parallel and congruent (the same). My angles are not 90°. Which shape am I? __________________	I have 4 sides. My top and bottom sides are parallel to each other. My right side and left side are not parallel to each other. Which shape am I? __________________
Mystery Shape 3	**Mystery Shape 4**
Only my opposite sides are parallel and congruent. All of my angles are 90°. Which shape am I? __________________	I have 4 sides that are congruent. My opposite sides are parallel to each other. All my angles are 90°. Which shape am I? __________________

Practice 33

Draw each shape and find the area. Use the formulas to help you.

Square	Rectangle
Area = Length x Width	Area = Length x Width
Width 2 units	Width 2 units
Length 2 units	Length 3 units
Area = 2 x 2	Area = 3 x 2
Area = 4 square units	Area = 6 square units

1. Ariel made a square. Each side is 4 units long. What is the square's area?

   The area is _____ square units.

2. Aaron made a rectangle with a length of 6 units and a width of 4 units. What is the rectangle's area?

   The area is _____ square units.

3. Sharon made a rectangle with a length of 5 units and a height of 3 units. What is the rectangle's area?

   The area is _____ square units.

4. Keith made a square. Each side is 5 units long. What is the square's area?

   The area is _____ square units.

Practice 34

Draw each triangle and find the area. Use the formula to help you.

$$\textbf{Area of Triangle} = \frac{\textbf{Base x Height}}{\textbf{2}}$$

Height = 4 units

Base = 3 units

$$\text{Area} = \frac{3 \times 4}{2}$$

$$\text{Area} = \frac{12}{2}$$

Area = 6 square units

1. Lola made a triangle with a base of 5 units and a height of 6 units. What is the triangle's area?

   The area is _____ square units.

2. Trey made a triangle with a base of 2 units and a height of 4 units. What is the triangle's area?

   The area is _____ square units.

3. Andy made a triangle with a base of 5 units and a height of 4 units. What is the triangle's area?

   The area is _____ square units.

4. Sue made a triangle with a base of 3 units and a height of 6 units. What is the triangle's area?

   The area is _____ square units.

Practice 35

Solve each word problem.

1. Evie is thinking of a number. The number is between 1 and 10. The number is divisible by 2 and by 3. What is Evie's number?

   Evie's number is _____.

2. Israel is thinking of a number. The number is between 11 and 20. The number is divisible by 2, 3, and 4. What is Israel's number?

   Israel's number is _____.

3. Charlotte is thinking of an odd number between 21 and 30. The number is divisible by 3 and by 9. What is Charlotte's number?

   Charlotte's number is _____.

4. Nathan is thinking of an even number between 31 and 40. The number is divisible by 2, 4, and 8. What is Nathan's number?

   Nathan's number is _____.

5. Kristen is thinking of an even number between 41 and 51. The number is divisible by 2, 5, 10, and 25. What is Kristen's number?

   Kristen's number is _____.

6. Lupe is thinking of an even number between 51 and 60. The number is divisible by 3, 6, and 9. What is Lupe's number?

   Lupe's number is _____.

Practice 36

To find the average, add all of the numbers in the group together and divide by the number of numerals in the group. Find the average length of each frog's hops.

1. Jumper jumped 3 cm. Froggy jumped 6 cm. Kicker jumped 9 cm. What was the length of the average jump?

   3 + 6 + 9 = 18

   18 ÷ 3 = __

   The average length was _____ cm.

2. Froggy jumped 4 cm. Jumper jumped 4 cm, too. Kicker jumped 7 cm, and Hopper jumped 9 cm. What was the length of the average jump?

   4 + 4 + 7 + 9 = __

   __ ÷ 4 = __

   The average length was _____ cm.

3. Jumper jumped 510 cm in 3 hops. What was the average length of one hop?

   $3\overline{)510}$

   The average length was _____ cm.

4. Froggy jumped 1,065 cm in 5 hops. What was the average length of one hop?

   $5\overline{)1065}$

   The average length was _____ cm.

5. Kicker hopped 888 cm in 4 hops. What was the average length of one hop?

   The average length was _____ cm.

6. Hopper hopped 460 cm in 10 hops. What was the average length of one hop?

   The average length was _____ cm.

Test Practice 1

Choose the correct answer.

1. Every Saturday afternoon, Nadine goes to see the latest movie playing at the Good Times Movie Theater. Today is Saturday. What is the chance that Nadine will go to see a movie this afternoon?

   (A) Likely (B) Unlikely

2. Shana has 18 necklaces, 19 bracelets, and 21 watches. About how many pieces of jewelry does Shana have? (Round each number to the nearest ten when solving.)

   (A) 20 (B) 40
   (C) 60 (D) 50

3. The distance from Craterville to Moon City is 167 miles. From Craterville to Sun City the distance is 639 miles. What is the estimated distance between Moon City and Sun City? (Round each number to the nearest hundred when solving the problem.)

   (A) 500 mi. (B) 300 mi.
   (C) 400 mi. (D) 200 mi.

4. Jose collected 22 seashells, 13 sand dollars, and 44 pebbles. How many items did Jose collect in all?

   (A) 79 (B) 78
   (C) 77 (D) 76

5. Raquel caught 87 fish. She threw 55 of them back into the ocean. How many fish does Raquel have left?

   (A) 22 (B) 142
   (C) 132 (D) 32

6. Lee takes a 15 minute nap every day and spends 1/2 of an hour exercising. How much time does Lee spend napping and exercising?

   (A) 30 minutes (B) 45 minutes
   (C) 60 minutes (D) 20 minutes

Test Practice 2

Choose the correct answer.

1. The horse trailer costs $3,076. The horse costs $4,803. What is the total cost for both items?

   Ⓐ $1,727 Ⓑ $1,879
   Ⓒ $7,879 Ⓓ $7,978

2. The newest computer game system regularly costs $5,444. Today it is on sale for $2,331. What is the total savings?

   Ⓐ $1,331 Ⓑ $3,311
   Ⓒ $7,775 Ⓓ $3,113

3. Sybil mixed together 1/3 cup of vinegar and 1/2 cup of oil. What is the total amount of liquid ingredients used?

   Ⓐ 1/5 Ⓑ 5/6
   Ⓒ 2/5 Ⓓ 6/5

4. The stadium holds 11,972 people. There are 9,081 fans at today's game. About how many empty seats are there? (Round each number to the nearest thousand.)

   Ⓐ 3,000 Ⓑ 2,000
   Ⓒ 4,000 Ⓓ 1,000

5. Penny the Party Planner needs 3,123 sets of silverware, 3,980 plates, and 1,572 napkins. How many party items does Penny need to get?

   Ⓐ 8,765 Ⓑ 8,567
   Ⓒ 8,675 Ⓓ 7,675

6. Fred had 8 small pouches. In each pouch Fred placed 9 small pebbles. How many pebbles does Fred have in all?

   Ⓐ 64 Ⓑ 74
   Ⓒ 17 Ⓓ 72

Test Practice 3

Choose the correct answer.

1. Lacy had $63.88. She spent $10.19 buying a new tire for her bike. How much money does Lacy have left?

   (A) $35.69 (B) $53.69
   (C) $55.69 (D) $56.39

2. Lance wants to buy a new bike seat for $4.56 and a new tire pump for $19.54. What is the total cost for both items?

   (A) $14.98 (B) $19.48
   (C) $24.10 (D) $21.40

3. Roma had $\frac{3}{4}$ of a dollar. She spent $0.18 buying a bike reflector. How much money does Roma have left?

   (A) $0.57 (B) $0.75
   (C) $0.93 (D) $0.32

4. John had half a dollar. He spent 23¢ buying a bike permit. How much money does John have left?

   (A) $0.28 (B) $0.27
   (C) $0.72 (D) $0.73

5. At the bike fair, 3,577 handlebar streamers and 6,098 bike locks were given away. What was the total number of items given away?

   (A) 2,521 (B) 9,765
   (C) 9,675 (D) 9,576

6. Out of the 3,326 visitors at the bike fair, 784 participated in the bike rodeo. How many people watched the bike rodeo?

   (A) 2,542 (B) 4,110
   (C) 2,452 (D) 4,100

Test Practice 4

Choose the correct answer.

1. Find the mystery number. The mystery number is an even number divisible by 3, 6, 11, and 33. What is the mystery number?

   (A) 61 (B) 63
   (C) 66 (D) 69

2. 8 cups of coffee can be made using 1 spoonful of coffee. If you need to make coffee for 80 people, how many spoonfuls of coffee will you need to use?

   (A) 10 (B) 9
   (C) 8 (D) 7

3. There are 36 pieces of licorice in a 6-ounce pack. How many pieces of licorice are in a 1-ounce pack?

   (A) 3 (B) 6
   (C) 9 (D) 1

4. The truck drove 500 miles on 10 gallons of gas. How many miles can the truck drive on one gallon of gas?

   (A) 5 (B) 15
   (C) 50 (D) 55

5. In the last 3 basketball games, Antonia scored 15, 36, and 27 points. What was Antonia's average score?

   (A) 26 (B) 28
   (C) 30 (D) 24

6. Ginny had 10 ribbons. She sold each ribbon for 9¢. How much money did Ginny earn?

   (A) 9¢ (B) 19¢
   (C) 99¢ (D) 90¢

Test Practice 5

Choose the correct answer.

1. Ramona bought a shovel for \$2.99, 3 packs of seeds for \$0.31 a pack, and 2 plant stakes for \$0.15 each. How much money did Ramona spend?

   (A) \$3.76 (B) \$4.07
   (C) \$4.22 (D) \$3.45

2. Martin bought 4 pairs of socks at 2 pairs for \$1.39, a belt for \$0.86, and a pair of pants for \$0.25. How much money did Martin spend?

   (A) \$3.89 (B) \$2.50
   (C) \$6.67 (D) \$3.98

3. Hilda bought 16 pencils to share with 4 friends. How many pencils can Hilda give to each friend?

   (A) 3 (B) 5
   (C) 4 (D) 2

4. Jackson bought 50 dog bones to feed to 5 dogs. How many dog bones can each dog have?

   (A) 5 (B) 20
   (C) 15 (D) 10

5. Kate had $\frac{1}{4}$ cup of pudding. She put $\frac{1}{6}$ of a cup of whipped cream on top. How much of a cup of pudding and whipped cream did Kate use in all?

   (A) 10/24 (B) 3/4
   (C) 2/10 (D) 1/2

6. Bill had $\frac{4}{5}$ of a cup of tuna fish. He used $\frac{1}{3}$ of a cup to make a tuna fish sandwich. How much tuna fish is left?

   (A) 5/8 (B) 3/2
   (C) 17/15 (D) 7/15

Test Practice 6

Choose the correct answer.

1. Armando had $4.00 in pennies. How many pennies did Armando have?

   Ⓐ 4 Ⓑ 40
   Ⓒ 4,000 Ⓓ 400

2. Andrea put 1,000 watermelon seeds into 100 small envelopes. How many seeds did Andrea put in each envelope?

   Ⓐ 1 Ⓑ 10
   Ⓒ 100 Ⓓ 1,000

3. A 6-pack of soda costs $2.04. What is the cost of one soda?

   Ⓐ $0.34 Ⓑ $0.35
   Ⓒ $0.33 Ⓓ $0.36

4. A 12-pack of gum costs $4.80. What is the cost of one stick of gum?

   Ⓐ $0.41 Ⓑ $0.39
   Ⓒ $0.04 Ⓓ $0.40

Bread	Muffins	Donuts
$2.40/loaf	$3.00/dozen	$2.40/dozen

5. Janell went to the bakery and bought 2 loaves of bread, 3 muffins, and 1 donut. How much did Janell spend?

   Ⓐ $2.33 Ⓑ $2.83 Ⓒ $7.78 Ⓓ $5.75

6. Jules bought half a dozen donuts and half a dozen muffins. How much did Jules spend?

   Ⓐ $2.70 Ⓑ $5.40 Ⓒ $4.20 Ⓓ $7.20

Answer Sheet

Test Practice 1 (page 40)	Test Practice 2 (page 41)	Test Practice 3 (page 42)
1. Ⓐ Ⓑ	1. Ⓐ Ⓑ Ⓒ Ⓓ	1. Ⓐ Ⓑ Ⓒ Ⓓ
2. Ⓐ Ⓑ Ⓒ Ⓓ	2. Ⓐ Ⓑ Ⓒ Ⓓ	2. Ⓐ Ⓑ Ⓒ Ⓓ
3. Ⓐ Ⓑ Ⓒ Ⓓ	3. Ⓐ Ⓑ Ⓒ Ⓓ	3. Ⓐ Ⓑ Ⓒ Ⓓ
4. Ⓐ Ⓑ Ⓒ Ⓓ	4. Ⓐ Ⓑ Ⓒ Ⓓ	4. Ⓐ Ⓑ Ⓒ Ⓓ
5. Ⓐ Ⓑ Ⓒ Ⓓ	5. Ⓐ Ⓑ Ⓒ Ⓓ	5. Ⓐ Ⓑ Ⓒ Ⓓ
6. Ⓐ Ⓑ Ⓒ Ⓓ	6. Ⓐ Ⓑ Ⓒ Ⓓ	6. Ⓐ Ⓑ Ⓒ Ⓓ

Test Practice 4 (page 43)	Test Practice 5 (page 44)	Test Practice 6 (page 45)
1. Ⓐ Ⓑ Ⓒ Ⓓ	1. Ⓐ Ⓑ Ⓒ Ⓓ	1. Ⓐ Ⓑ Ⓒ Ⓓ
2. Ⓐ Ⓑ Ⓒ Ⓓ	2. Ⓐ Ⓑ Ⓒ Ⓓ	2. Ⓐ Ⓑ Ⓒ Ⓓ
3. Ⓐ Ⓑ Ⓒ Ⓓ	3. Ⓐ Ⓑ Ⓒ Ⓓ	3. Ⓐ Ⓑ Ⓒ Ⓓ
4. Ⓐ Ⓑ Ⓒ Ⓓ	4. Ⓐ Ⓑ Ⓒ Ⓓ	4. Ⓐ Ⓑ Ⓒ Ⓓ
5. Ⓐ Ⓑ Ⓒ Ⓓ	5. Ⓐ Ⓑ Ⓒ Ⓓ	5. Ⓐ Ⓑ Ⓒ Ⓓ
6. Ⓐ Ⓑ Ⓒ Ⓓ	6. Ⓐ Ⓑ Ⓒ Ⓓ	6. Ⓐ Ⓑ Ⓒ Ⓓ

Answer Key

Page 4

1. 60 4. 20
2. 80 5. 90
3. 30 6. 30

Page 5

Chart numbers

100, 100, 100
100, 200, 200
100, 200, 200
100, 200, 200

1. 200
2. 300
3. 500

Page 6

1. 442
2. 666
3. 413
4. 322
5. 524
6. 557

Page 7

1. 7,130
2. 6,488
3. 7,717
4. 3,988
5. 3,232
6. 5,349

Page 8

1. 73
2. 73
3. 77
4. 68
5. 66,757
6. 45,883

Page 9

1. 81
2. 89
3. 78
4. 192
5. 2833
6. 1,417

Page 10

1. 146,909
2. 2,370
3. 19,502
4. 9,085

Page 11

1. 8
2. 4
3. 2 x 3 = 6
4. 4 x 4 = 16
5. 8 x 8 = 64
6. 5 x 8 = 40

Page 12

1. 5 x 1¢ = 5¢
2. 8 x 3¢ = 24¢
3. 6 x 4 = 24
4. 3 x 3 = 9
5. 5 x 6¢ = 30¢
6. 3 x 9 = 27

Page 13

1. 48 eggs
2. 48 hours
3. 48 months
4. 77 days
5. 365 days
6. 210 days
7. 60 inches
8. 54 feet

Page 14

1. 8 beads
2. 4 seeds
3. 8 legs
4. 2 wings
5. 5 feathers
6. 2 ears

Page 15

1. 48 ÷ 4 = 12
   12 x 4 = 48
2. 72 ÷ 2 = 36
   36 x 2 = 72
3. 39 ÷ 3 = 13
   13 x 3 = 39
4. 44 ÷ 4 = 11
   11 x 4 = 44

Page 16

1. 15
2. 90
3. 50
4. 70
5. 25
6. 11
7. 2
8. 8

Page 17

1. 230
2. 15
3. 8 lbs.
4. 400 hours
5. 100
6. 200

Page 18

1. 1,000
2. 18
3. $25.00
4. 50
5. 2,000
6. $10.00

Page 19

1. 24 R2
2. 6 R2
3. 8 R3
4. 9 R2
5. 23 R5
6. 97 R1

Page 20

1. 2,635
2. 827
3. 4,355
4. 10,584; 7,938

Page 21

1. 23 min.
2. 45 min.
3. 35 min.
4. 45 min.
5. 25 min.
6. 39 min.

Page 22

1. 2 hours, 15 minutes
2. 1 hour, 45 minutes
3. 2:15
4. 6:45
5. 2:47
6. 52 minutes

Page 23

1. $1.87
2. $1.32
3. $2.22
4. $2.34

Page 24

1. 41¢
2. 36¢
3. 49¢
4. 57¢
5. 29¢
6. 3¢
7. Louanne
8. Matilda

Page 25

1. $78.91 – $78.45 = $0.46
2. $45.52 – $13.61 = $31.91
3. $12.63 – $11.09 = $1.54
4. $92.99 – $56.93 = $36.06

Page 26

1. $40.46
2. $36.19
3. $11.09
4. $36.57
5. $36.82
6. $45.45

Answer Key (cont.)

Page 27

1. $2.27
2. $2.64
3. $1.02
4. $1.57
5. $0.61
6. $0.57
7. $0.41
8. $1.48

Page 28

1. 1
2. 6
3. 3
4. 7
5. 8
6. 6

Page 29

1. 3/6 or 1/2
2. 4/6 or 2/3
3. 1/8
4. 2/3
5. 3/4
6. 5/6

Page 30

1. 4/9
2. 4/8 or 1/2
3. 5/7
4. 6/11
5. 3/5
6. 4/10 or 2/5

Page 31

1. 3/4
2. 4/5
3. 2/4 or 1/2
4. 2/6 or 1/3
5. 3/8
6. 12/24 or 1/2

Page 32

1. 7/12
2. 13/30
3. 35/50 or 7/10
4. 8/15
5. 19/15 or 1 4/15
6. 1/20

Page 33

1. Likely
2. Unlikely
3. Likely
4. Unlikely
5. Unlikely
6. Likely

Page 34

1. About 50
2. > 35
3. > 100
4. > 110
5. < 18
6. > 80

Page 35

1. rhombus
2. trapezoid
3. rectangle
4. square

Page 36

1. 16
2. 24
3. 15
4. 25

Page 37

1. 15
2. 4
3. 10
4. 9

Page 38

1. 6
2. 12
3. 27
4. 32
5. 50
6. 54

Page 39

1. 6 cm
2. 6 cm
3. 170 cm
4. 213 cm
5. 222 cm
6. 46 cm

Page 40

1. ● Ⓑ
2. Ⓐ Ⓑ ● Ⓓ
3. Ⓐ Ⓑ ● Ⓓ
4. ● Ⓑ Ⓒ Ⓓ
5. Ⓐ Ⓑ Ⓒ ●
6. Ⓐ ● Ⓒ Ⓓ

Page 41

1. Ⓐ Ⓑ ● Ⓓ
2. Ⓐ Ⓑ Ⓒ ●
3. Ⓐ ● Ⓒ Ⓓ
4. ● Ⓑ Ⓒ Ⓓ
5. Ⓐ Ⓑ ● Ⓓ
6. Ⓐ Ⓑ Ⓒ ●

Page 42

1. Ⓐ ● Ⓒ Ⓓ
2. Ⓐ Ⓑ ● Ⓓ
3. ● Ⓑ Ⓒ Ⓓ
4. Ⓐ ● Ⓒ Ⓓ
5. Ⓐ Ⓑ ● Ⓓ
6. ● Ⓑ Ⓒ Ⓓ

Page 43

1. Ⓐ Ⓑ ● Ⓓ
2. ● Ⓑ Ⓒ Ⓓ
3. Ⓐ ● Ⓒ Ⓓ
4. Ⓐ Ⓑ ● Ⓓ
5. ● Ⓑ Ⓒ Ⓓ
6. Ⓐ Ⓑ Ⓒ ●

Page 44

1. Ⓐ Ⓑ ● Ⓓ
2. ● Ⓑ Ⓒ Ⓓ
3. Ⓐ Ⓑ ● Ⓓ
4. Ⓐ Ⓑ Ⓒ ●
5. ● Ⓑ Ⓒ Ⓓ
6. Ⓐ Ⓑ Ⓒ ●

Page 45

1. Ⓐ Ⓑ Ⓒ ●
2. Ⓐ ● Ⓒ Ⓓ
3. ● Ⓑ Ⓒ Ⓓ
4. Ⓐ Ⓑ Ⓒ ●
5. Ⓐ Ⓑ Ⓒ ●
6. ● Ⓑ Ⓒ Ⓓ